Upgrade Your Ph

R. H. C. Neill
G. Sydserff

EDWARD ARNOLD

Authors' preface

The authors feel that many pupils find considerable difficulty in relating the problems which they are required to solve to the underlying theory. With a packed syllabus to get through, the teachers, however willing, cannot always find the time to develop the pupils' examination techniques as well as they would like to. *Upgrade Your Physics* has been designed to help to overcome these difficulties with regard to senior school physics courses, in the same way that our earlier publication *Physics: What's the Problem?* was developed to assist the O-Grade/Level courses.

Upgrade Your Physics is based upon the Scottish Higher Grade syllabus, but the material covered should be useful for a wide variety of senior school examinations. The book is presented in such a way that it can be used by pupils who want an effective method of revising Higher Grade/Senior Grade physics for examinations, or by pupils who want to improve their problem-solving techniques.

In the introductory section on 'How to use this book' we have stressed that the book has been structured to cater for different individual needs. There are six units in the text covering the principal subject areas encountered in most senior school physics courses (Ray optics; Wave motion; Mechanics; Heat and kinetic theory of gases; Electricity; Atomic physics). Within each unit, the topics are covered by a sequence of examples to which *fully worked solutions* are provided. The examples and their solutions are printed in separate sections of the book so that the reader can study the material in a variety of ways.

SI units have been used throughout. The worked examples have been given a mark allocation, but this is only included for guidance.

We hope that this new book, *Upgrade Your Physics*, will appeal to teachers searching for problems and to pupils searching for answers.

Robert H. C. Neill
George Sydserff
Edinburgh, June, 1980

Acknowledgements

We are once more indebted to our artist Tony Merriman. Never, in the history of physics problem books, have so many drawings been produced in so short a time to such great effect.

Contents

		page
How to use this book		iv
Ray optics	worked examples	1
	solutions to worked examples	5
	practice questions	9
Wave motion	worked examples	11
	solutions to worked examples	13
	practice questions	15
Mechanics	worked examples	16
	solutions to worked examples	23
	practice questions	32
Heat and kinetic theory of gases	worked examples	36
	solutions to worked examples	39
	practice questions	42
Electricity	worked examples	44
	solutions to worked examples	60
	practice questions	74
Atomic physics	worked examples	82
	solutions to worked examples	84
	practice questions	86
Answers to practice questions		87
Index to worked examples		90

How to use this book

How this book is set out

There are six **units** in this book, covering the six major topics of most senior school physics courses :

unit L	on **Ray optics**
unit W	on **Wave motion**
unit M	on **Mechanics**
unit H	on **Heat and kinetic theory of gases**
unit E	on **Electricity**
unit A	on **Atomic physics**.

Each unit has two main parts :
(a) A set of questions presented in an acceptable teaching order.
(b) A corresponding set of detailed solutions to these questions.
In addition, each unit has a batch of practice problems which are modelled on the above worked examples. Practice problem numbers are followed by a bracketed label showing which worked examples they refer to. Answers only are supplied for these problems and these are given at the end of the book.

There are several ways of using this book

- If you want a way to revise your physics course *quickly* for an examination, read each example in conjunction with its solution and work your way steadily through the units.
- If you want to practise solving physics questions, try to do each question yourself without looking at its solution. Then check your answer carefully against the solution provided.
- If you want to study questions on a particular topic (e.g. Refractive Index) look up this topic in the index at the back of the book to find a list of all the relevant examples. Then study these carefully.

Remember to try the practice problems. They will test your progress.

UNIT L—Ray optics

Worked Examples

L1 Snell's law of refraction
Here is an extract from a pupil's laboratory notebook, showing results of what happens to a ray of light when it goes from air into glass.

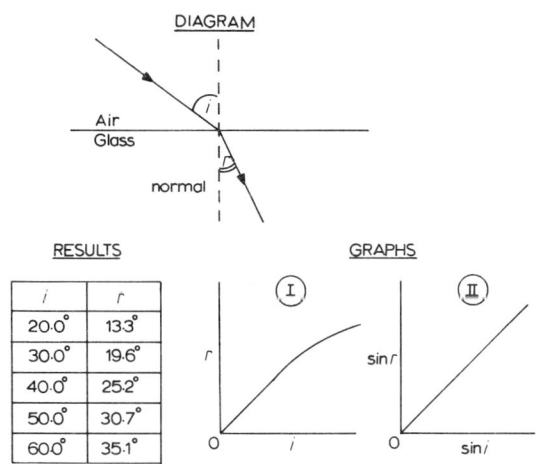

i	r
20.0°	13.3°
30.0°	19.6°
40.0°	25.2°
50.0°	30.7°
60.0°	35.1°

(a) What name is given to the effect of light bending when it passes from one medium into another? (1)
(b) Angle i is called the angle of incidence. What name is given to angle r? (1)
(c) What famous law of optics is the subject of this experiment? (1)
(d) Explain the significance of the form of the two graphs. (3)
(e) Calculate the average value for the refractive index of glass relative to air. (2)
(f) If, instead of glass, a material with a higher refractive index were used, what would be the effect upon Graph II? (2)

L2 Refractive index
In the following table, the refractive index (n) is given for a number of substances.

Substance	glass	water	ice	diamond
Refractive index	1.5–1.9	1.33	1.31	2.42

(a) Give a possible reason for glass having a range of values for its refractive index. (2)
(b) Assuming the value 30° for the angle of incidence, calculate the angle of refraction for (i) water and (ii) ice. (4)
(c) A ray of light passes from air into diamond, making a refracted angle of 15°. What is the angle of incidence of the ray of light approaching the diamond? (2)

2 Ray optics

(d) The refractive index of red light is 1.513 for light going from air into crown glass, whereas for violet light it is 1.532. Use this information to help you to explain why a spectrum can be produced when white light passes through a crown glass prism. (2)

L3 Critical angle and total internal reflection

A ray of light is shown entering a semicircular block of glass at P and then leaving it at O, which is the mid-point of the straight edge.

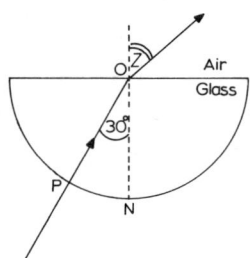

(a) Why is there no bending of the ray of light when it enters the glass? (2)
(b) Taking 30° as the angle of incidence and z as the angle of refraction, write down the relationship between these angles and the refractive index of the glass relative to air. (Value of $_{air}n_{glass}$ = 1.5) (2)
(c) Calculate the value of angle z. (2)
(d) Angle PON is increased until angle z becomes 90°. Calculate the value of angle PON. (2)
(e) What is the special name given for angle PON when angle z = 90°? Explain its significance. (2)

L4 Ray tracing

A pupil sets up a neon laser and directs the laser beam parallel to the axis of a semi-circular glass block. Two different positions of the block are shown below. The refractive index of the glass relative to air is known to be 1.5 for this block.

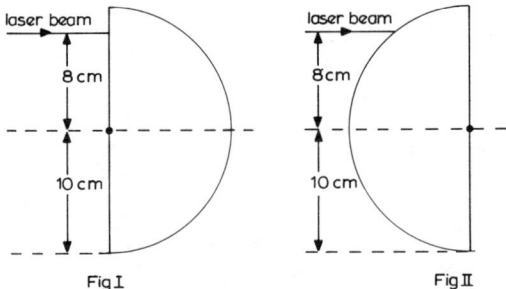

The pupil discovers that the laser beam follows quite different paths for the two positions of the block.

(a) Explain why, for Fig. I, the laser beam makes two internal reflections before emerging from the block. Produce a rough sketch of the situation. (4)
(b) Trace the path of the laser beam through the block for Fig. II, by drawing a scale diagram and by marking angles in where necessary. (6)

L5 Image formation by a lens

In the table below, the row referring to the image shown on the following diagram has been completed for you. When an object is placed between positions 2F' and F' (see diagram), its image could be described as **real, inverted** and **magnified**.

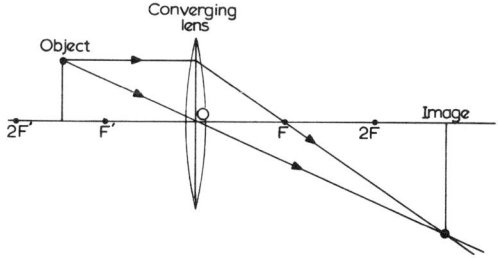

(a) Complete the rest of this table to describe the images of the object when placed in other positions. (9)

Object position	Description of image						
	Real	Virtual	Upright	Inverted	Magnified	Diminished	Same size
Between infinity and 2F'							
At 2F'							
Between 2F' and F'	√	—	—	√	√	—	—
Between F' and lens							

(b) What happens in the special case when the object is placed at F'? (1)

L6 Introducing the lens equation

Rays of light from a point A on an object are refracted by a thin lens and they come together at point D. Two of these rays are shown below.

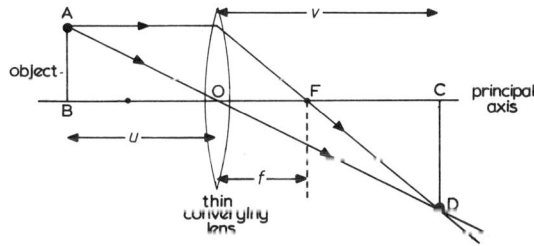

(a) Describe the paths taken by the two rays which have been used to locate point D. (2)

4 Ray optics

(b) Where do all the light rays, which leave point B and pass through the lens, converge? (1)
(c) If line AB represents an object, what does line CD represent? (1)
(d) A pupil places a match 15 cm away from a thin converging lens of focal length $f = 10$ cm. Draw a scale diagram to find the image distance (v) for this match. Describe the image obtained. (4)
(e) The relationship $\dfrac{1}{u} + \dfrac{1}{v} = \dfrac{1}{f}$ can be used to find the value of the image distance (v), provided the object distance (u) and the focal length of the lens (f) are both known. Use this lens equation to calculate the value of v and compare it with the result found in (d). (2)

L7 Production of magnified images

A photographic enlarger incorporates a bellows assembly so that the lens to negative distance can be altered. The enlarger shown has a converging lens of focal length 6 cm.

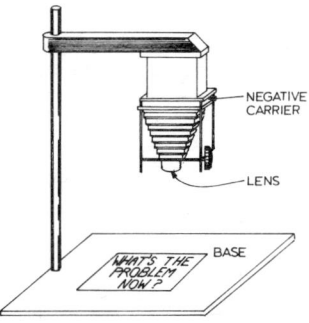

A negative measuring 3.5 cm by 2.5 cm is placed in the carrier and the lens position is adjusted until a sharp image is obtained when the lens is 42 cm above the base.
(a) How far is the negative away from the lens? (2)
(b) Describe the image produced. (3)
(c) What size is the image produced by the enlarger? (2)
(d) What adjustment would you make to the enlarger to produce a *bigger, in-focus* image on the base plate? (3)

L8 The simple telescope

Here is a diagram of a simple astronomical telescope:

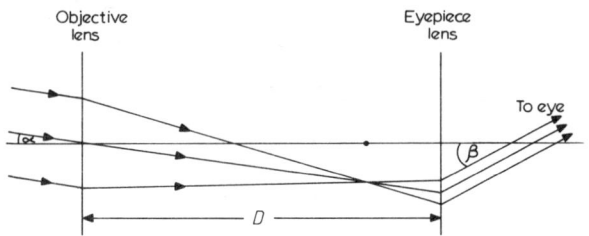

(a) If the rays of light entering the objective lens are parallel, and if the rays of light emerging from the eyepiece lens are also parallel, what can you say about the distance D which separates the two lenses? (1)
(b) How does the objective lens compare with the eyepiece lens in terms of
  (i) lens diameter?
  (ii) focal length? (2)
(c) Explain how magnification arises from this combination of lenses. (3)
(d) Describe how the telescope could be adapted to be used as a terrestrial telescope. (4)

Solutions to Worked Examples

L1

(a) Refraction.

(b) Angle r is called the angle of refraction. When light passes from air into glass, the light bends towards the normal and so angle r is less than angle i.

(c) The law being investigated is Snell's Law which states that $\dfrac{\sin i}{\sin r}$ = constant (refractive index).

(d) Graph I shows that for small angles of incidence $\dfrac{i}{r}$ = constant. However, as i increases, the ratio $\dfrac{i}{r}$ is no longer constant (graph curves). Graph II shows that, for all values of i, the ratio $\dfrac{\sin i}{\sin r}$ = constant.

(e)

i	r	$\sin i$	$\sin r$	$\dfrac{\sin i}{\sin r}$
20.0°	13.3°	0.342	0.230	1.49
30.0°	19.6°	0.500	0.335	1.49
40.0°	25.2°	0.643	0.426	1.51
50.0°	30.7°	0.766	0.511	1.50
60.0°	35.1°	0.866	0.575	1.51

From the last column, the average value of $\dfrac{\sin i}{\sin r}$ is 1.50

(f) For any selected value of i, the corresponding value for r would be *smaller*, indicating more bending towards the normal. Thus, for any selected value of $\sin i$, the value of $\sin r$ would be *smaller*, making the gradient of the graph *less steep*.

L2

(a) The reason behind there being many different values for n_{glass} is the fact that there are many types of glass with different optical properties, e.g. crown glass, flint glass, soda glass, etc.

6   Ray optics

(b) Snell's Law is used here : $\dfrac{\sin i}{\sin r} = n$.

   (i) $\dfrac{\sin 30}{\sin r} = n_{water} = 1.33 \Rightarrow \sin r = \dfrac{0.500}{1.33} = 0.376 \Rightarrow r = 22.1°$

   (ii) $\dfrac{\sin 30}{\sin r} = n_{ice} = 1.31 \Rightarrow \sin r = \dfrac{0.500}{1.31} = 0.382 \Rightarrow r = 22.4°$

(c) $\dfrac{\sin i}{\sin 15} = n_{diamond} = 2.42 \Rightarrow \sin i = 2.42 \times 0.259 = 0.626 \Rightarrow i = 38.8°$

(d) The refractive indices for red and violet light are different. This means that beams of red and violet light having the same angle of incidence on a glass prism will be refracted at different angles. Since these colours form the limits of the spectrum, the difference between their refracted angles will contain the entire spectrum.

L3

(a) Here the ray of light is entering the block of glass along the normal, in this case along a radius. There is therefore no bending. Refraction here amounts to a reduction in speed and wavelength.

(b) $\dfrac{\sin i}{\sin r} = \dfrac{\sin 30}{\sin z} = \dfrac{1}{_{air}n_{glass}} = \dfrac{1}{1.5}$

Notice that the ratio of the sines is less than 1 because 30° is less than z. Remember that, if $_{air}n_{glass} = 1.5$, then $_{glass}n_{air} = \dfrac{1}{1.5}$

(c) By Snell's Law of Refraction

$\dfrac{\sin 30}{\sin z} = \dfrac{1}{1.5} \Rightarrow \sin z = 1.5 \sin 30 \Rightarrow \sin z = 1.5 \times 0.500 = 0.750$

From natural sine tables, $z = 48.6°$.

(d) Here $\dfrac{\sin PON}{\sin z} = \dfrac{\sin PON}{\sin 90} = \dfrac{1}{1.5}$. But $\sin 90 = 1 \therefore \sin PON = \dfrac{1}{1.5}$

$\sin PON = 0.667 \Rightarrow$ angle $PON = 41.8°$

(e) This special angle of incidence which produces a refracted angle of 90° is called the **critical angle**. At angles greater than this critical angle all the light arriving at O is totally internally reflected. No light passes out from the glass into air at O.

L4

(a) The laser beam in Fig. I continues with no change in direction until it meets the curved surface of the glass block. The angle of incidence which it makes with the curved surface is given by $\sin i = \dfrac{8\ cm}{10\ cm} = 0.8$ and so $i = 53.1°$.

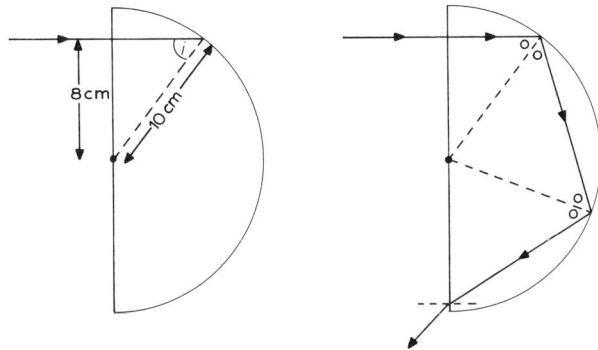

This angle is greater than the critical angle which is 41.8° (see L3) and so the laser beam will be totally internally reflected.

(b)

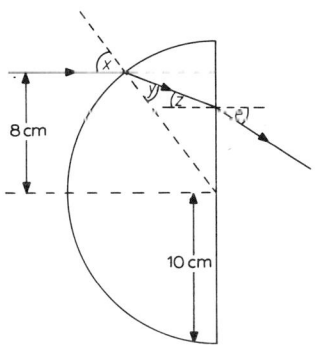

$$\sin x = \frac{8 \text{ cm}}{10 \text{ cm}} = 0.8 \Rightarrow x = 53.1°$$

$$\frac{\sin x}{\sin y} = 1.5 \Rightarrow \sin y = \frac{\sin x}{1.5} = \frac{0.8}{1.5} = 0.5333 \Rightarrow y = 32.2°$$

Angle z = angle x − angle y = 53.1° − 32.2° = 20.9°

$$\frac{\sin z}{\sin e} = \frac{1}{1.5} \Rightarrow \sin e = 1.5 \sin z = 1.5 \sin 20.9 = 0.5351 \Rightarrow \text{angle } e = 32.4°$$

L5
(a) *Object between ∞ and 2F'* : **real, inverted, diminished.**
Object at 2F' : **real, inverted, same size.**
Object between 2F' and F' : **real, inverted, magnified.**
Object between F' and lens : **virtual, upright, magnified.**
(b) When the object is placed at F' no image is formed. The two rays usually used to locate images emerge from the lens parallel to each other.

8   Ray optics

L6

(a) Two rays are needed to locate the image of point A on the object.
    Ray I is drawn parallel to the principal axis of the lens and is then refracted so that it passes through the focus F.
    Ray II passes through the optical centre of the lens and remains undeviated, provided the lens is very thin.
    Rays I and II converge and meet at point D, helping to form the image of A at D. In fact *all* other rays starting from A and passing through this lens would meet at D.
(b) These rays would converge at point C on the principal axis.
(c) Line CD is called the linear image of object BA.
(d) Using a suitable scale drawing, the image is located about 30 cm from the lens on the opposite side of the lens from the object. The image is real, inverted and magnified, being about twice as large as the object.
    When making a ray diagram, remember to quote the scale you have decided to use and remember to put arrows on the rays to indicate the direction of the energy.
(e) Here u = distance of object from lens centre O = 15 cm.
    $\quad v$ = distance of image from lens centre O = ?
    $\quad f$ = focal length of the lens = 10 cm.
    Substituting these figures in the lens equation, we get

$$\frac{1}{u} + \frac{1}{v} = \frac{1}{f} \Rightarrow \frac{1}{v} = \frac{1}{f} - \frac{1}{u} = \frac{1}{10} - \frac{1}{15} = \frac{1}{30}$$

Inverting both sides, we see that the image distance v = 30 cm. Also, we can obtain the magnification M as follows:

$$\text{linear magnification} = M = \frac{v}{u} = \frac{30}{15} = 2. \text{ This agrees with (d).}$$

L7

(a) The image distance v is 42 cm and the focal length is 6 cm. We can find the required object distance u from the Lens Equation.

$$\frac{1}{u} + \frac{1}{v} = \frac{1}{f} \Rightarrow \frac{1}{u} = \frac{1}{f} - \frac{1}{v} = \frac{1}{6} - \frac{1}{42} = \frac{6}{42} = \frac{1}{7}$$

Inverting, we see that the required object distance u = 7 cm.
The lens is 7 cm away from the negative.

(b) The image is **real, inverted** (usually the negative is inverted to compensate for this) and **magnified**.

(c) The size of the image is obtained from the formula for Linear Magnification.

$$M = \frac{v}{u} = \frac{42 \text{ cm}}{7 \text{ cm}} = 6.$$ Since the negative is magnified six times, its image measures 21 cm by 15 cm.

(d) The enlarger would be raised to increase the distance of the lens from the base. The lens would then be moved closer to the negative until the new, bigger image came into focus.

L8

(a) The distance between the two lenses would have to equal the sum of their focal lengths:

$$D = f_{\text{objective}} + f_{\text{eyepiece}}$$

(b) (i) The objective lens has a larger diameter to admit more light to the telescope.
(ii) The objective lens also has a longer focal length.
(c) Originally, the parallel rays of light coming from a point on a distant object subtend a very small angle α to the principal axis.
After passing through the telescope, the rays emerge at a much greater angle β. Thus light enters the eye at a much larger angle *with* the telescope, than it would *without* the telescope. This produces a larger image on the retina of the eye, i.e. things look nearer.
(d) A third, erecting lens is incorporated into the astronomical telescope between the objective and the eyepiece. This extra lens produces an inversion of the image obtained. Instead of seeing things upside down (as with the astronomical telescope), things are seen the right way up with this terrestrial telescope.
The length of this new telescope is actually given by

$$D = f_{objective} + f_{eyepiece} + 4 \times f_{erecting\ lens}$$

Read an advanced textbook on optics for a more detailed description!

Practice Questions

1. (L1)

Snell's Law of Refraction can be written in the form $\dfrac{\sin i}{\sin r}$ = constant.

(a) What name is given to the constant?
(b) What affects the value of this constant?
(c) A ray of light is incident upon a block of transparent material at 45°. The angle of refraction is 30°. Calculate the value of the constant in this case.

2. (L2)

The refractive index of glass relative to air is 1.5.
(a) A ray of light in air makes an angle of incidence of 60° with a glass block. Calculate the corresponding angle of refraction.
(b) Using the same block of glass, the ray in air is altered so that the angle of refraction in glass is 30°. Calculate the corresponding angle of incidence.
(c) When light passes from air into a liquid, the angles of incidence and refraction are 48.0° and 37.0° respectively. Calculate the refractive index of the liquid relative to air.

3. (L2)

The refractive index of red light is 1.513 for light going from air into crown glass whereas for violet light it is 1.532. When a ray of white light travels from air into crown glass at an angle of incidence of 55.0°, a small cone of colour is produced by dispersion. Calculate the apex angle of this cone.

4. (L3)

The refractive indices of water and diamond are 1.33 and 2.42 respectively.
(a) Calculate the corresponding critical angles for light travelling from each substance into air.
(b) Explain why light striking the back surface of a glass prism at 45° is totally internally reflected.

10 Ray optics

5. (L4)

A monochromatic ray of light travelling in air meets a prism made of a type of glass whose refractive index is 1.50 relative to air. The sides of the prism are all 10 cm long and the light hits the mid-point of one side with an angle of incidence of 23°. Trace the path of the ray through the prism until it re-emerges into air.

6. (L5)

The power of a lens is given by $P = \dfrac{1}{f}$, where f is the focal length expressed in metres, and P is the power in dioptres.
(a) Show by drawing a ray diagram that a lens of power 10 dioptres will form a real, inverted and magnified image of an object placed 15 cm from the lens.
(b) How far is this image from the lens?

7. (L6)

A magnifying glass placed 5 cm directly above any letter on this page magnifies the letter 8 times and produces an upright image of the letter.
(a) What is the focal length of the magnifying lens?
(b) How would you describe the image produced?

8. (L7)

The average width of a frame of movie film is 5 mm. A projectionist adjusts the projector to give a sharp picture which fills a screen 2 m wide when the screen is 12 m away from the lens of the projector.
(a) Find (i) the distance of the film from the lens, (ii) the focal length of the lens.
(b) The screen is now brought closer to the projector. What adjustment to the lens would have to be made to give a sharp (in focus) image on the screen? What can you say about the size of the image?

UNIT W—Wave motion

Worked Examples

W1 The wave theory of refraction

Plane water waves are produced by a spar vibrating at 5 Hz. They travel from a deep to a shallow region of a ripple tank as shown.

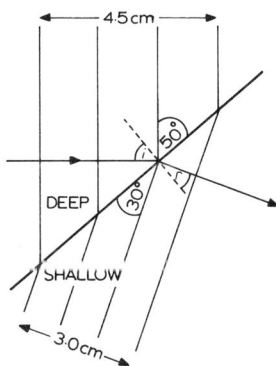

(a) From the measurements shown, calculate the wavelength, frequency and speed of the water waves in the *shallow region*. (3)

(b) Show that the value of $\dfrac{\sin i}{\sin r}$ is approximately equal to $\dfrac{\lambda_{deep}}{\lambda_{shallow}}$. (3)

(c) In your study of light, you will have come across the relationships

refractive index, $n = \dfrac{\sin i}{\sin r} = \dfrac{\lambda_1}{\lambda_2} = \dfrac{v_1}{v_2}$.

Using a ray of light of frequency 5.0×10^{14} Hz, in air, and taking the refractive index of diamond, relative to air, to be $n = 2.42$, calculate
 (i) the frequency of the light in diamond,
 (ii) the wavelength of the light in air,
 (iii) the velocity of the light in diamond.
  (Velocity of light in air $= 3.0 \times 10^8$ m s^{-1}). (4)

W2 Interference of light waves

The diagram represents the apparatus used in Young's double-slit experiment on the interference of light. S is a source of monochromatic light. The two slits S_1 and S_2 were made with a razor blade on a sheet of blackened glass. The separation of the slits is d metres and the light falls on a white screen which is set up D metres away from the slits. Distance d is very much smaller than distance D.

12 Wave motion

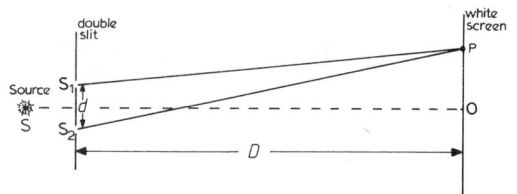

(a) Describe the interference pattern obtained on the white screen. (2)
(b) What condition must apply to the point P so that it is a position of maximum light intensity? *Hint*: Consider the path difference ($S_2P - S_1P$). (2)
(c) The distance between two neighbouring fringes on the screen is given by

$$\Delta y = \lambda \cdot \frac{D}{d}$$

(i) How would you measure Δy and d in an actual experiment? (3)
(ii) The second minimum position from O is found to be 0.6 cm along the screen from O. What is the value of Δy? (1)
(iii) Calculate the wavelength of the source if the ratio $\frac{D}{d}$ = 6780. (2)

W3 Interference of sound waves

Two loudspeakers are connected in parallel to the output of a signal generator which has been set to a frequency of 3.30 kHz. A small, sensitive microphone is moved along a scale AOB which has its zero at O. The output of the microphone is displayed on an oscilloscope.

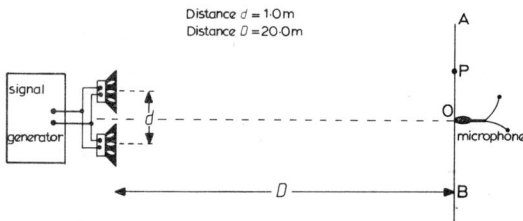

A graph of the output voltage of the microphone at various positions along the scale is shown below. The distances are measured in metres.

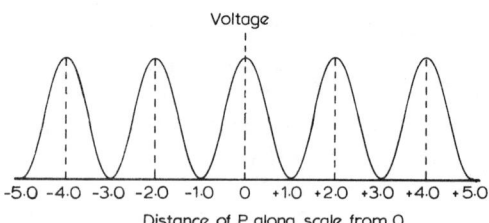

(a) What is taking place at position O to account for a maximum signal there? (1)
(b) What is happening at point P on the scale when
    (i) P is 1.0 m from O? (ii) P is 4.0 m from O? (4)
(c) In the study of interference, Δy is used to represent the distance between neighbouring maxima or minima along the scale.
    (i) What is the value of Δy here?

(ii) If d represents the distance between the loudspeakers and D represents the distance from the loudspeakers to the scale, show that

$$\Delta y = \lambda \cdot \frac{D}{d}$$

by using the values given in this experiment.
(iii) What would be the effect upon the separation of maxima of doubling the frequency of the signal generator?
(iv) What would be the effect upon the separation of maxima of doubling the distance between the loudspeakers? (5)
(Velocity of sound in air = 330 m s^{-1})

W4 The spectrometer
The diagram represents a typical spectrometer arrangement being used to look at light from a sodium lamp.

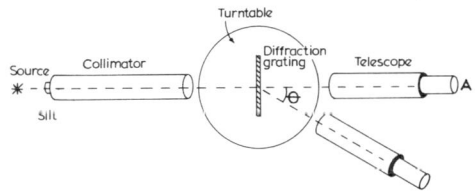

(a) Describe briefly the procedure used for setting up a spectrometer, before inserting the diffraction grating, so that a focused image of the collimator slit is obtained. (2)
(b) What will be seen in position A of the telescope when the diffraction grating is inserted as shown? (2)
(c) The telescope is slowly rotated from position A, increasing angle θ as shown until a first order spectrum of the source is seen. Contrast this with what is seen at position A. (2)
(d) According to the theory of the diffraction grating, $\sin \theta = \dfrac{n\lambda}{d}$

What do the symbols n, λ and d stand for? (2)
(e) If the sodium source is replaced by a mirror reflecting sunlight towards the collimator slit, describe what will be seen as the telescope is rotated from position O. (2)

Solutions to Worked Examples

W1

(a) $\lambda_{shallow} = \dfrac{3.0 \text{ cm}}{3} = 1.0 \text{ cm}, \quad f_{shallow} = f_{deep} = 5 \text{ Hz}$

$v_{shallow} = f_{shallow} \lambda_{shallow} = 5 \times 1.0 = 5 \text{ cm s}^{-1}$

(b) $\dfrac{\sin i}{\sin r} = \dfrac{\sin 50}{\sin 30} = 1.53, \quad \dfrac{\lambda_{deep}}{\lambda_{shallow}} = \dfrac{1.5 \text{ cm}}{1.0 \text{ cm}} = 1.5.$ Ratios almost equal.

(c) (i) There is no change in frequency so $f_{diamond} = 5.0 \times 10^{14}$ Hz.

(ii) $\lambda_{air} = \dfrac{v_{air}}{f} = \dfrac{3.0 \times 10^8 \text{ m s}^{-1}}{5.0 \times 10^{14} \text{ Hz}} = 6.0 \times 10^{-7} \text{ m}$

14 Wave motion

(iii) $\dfrac{v_{air}}{v_{diamond}} = n = 2.42 \Rightarrow v_{diamond} = 1.2 \times 10^8$ m s^{-1}.

W2
(a) A set of interference fringes would appear on the white screen, made up of dark and bright lines parallel to the slits and almost equally spaced. The central bright line in the pattern would be at position O.
(b) For maximum light intensity, the path difference ($S_2P - S_1P$) has to equal a whole number of wavelengths of the light being used.
(c) (i) *To measure Δy* : Use a ruler to find the total width of as many fringes as can be clearly seen on the white screen. Divide by this number to establish the width of one fringe (Δy).
To measure d : Clamp the blackened glass plate under the carriage of a travelling microscope and centre the microscope above one slit. Then move microscope along until it centres upon the other slit. Read the separation distance from the vernier scale on the travelling microscope.

(ii) $\Delta y = \dfrac{2}{3} \times 0.6$ cm $= 0.4$ cm

(iii) $\lambda = \Delta y \cdot \dfrac{d}{D} = 0.4 \times 10^{-2}$ m $\times \left(\dfrac{1}{6780}\right) = 5.9 \times 10^{-7}$ m

W3
(a) At O, the sound waves from the two loudspeakers interfere constructively to give a maximum signal. The path lengths of the two sets of sound waves from loudspeaker up to O are equal and so the waves arrive at O **in phase**.
(b) (i) When P is 1.0 m from O, the signal is zero. This indicates that

the waves from the two loudspeakers are arriving **out of phase** ;

the path difference between the two sets of waves is $\dfrac{\lambda}{2}$;

destructive interference is taking place.

(ii) When P is 4.0 m from O, the signal is again a maximum. This indicates that

the waves from the two loudspeakers are arriving **in phase** ;
the path difference between the two sets of waves is 2λ ;
constructive interference is taking place.
(c) (i) $\Delta y = 2.0$ m (separation of maxima or minima on graph)
(ii) $\Delta y = 2.0$ m

$$\lambda \cdot \dfrac{D}{d} = \dfrac{v}{f} \cdot \dfrac{D}{d} = \dfrac{330 \text{ m s}^{-1} \times 20.0 \text{ m}}{3300 \text{ Hz} \times 1.0 \text{ m}} = 2.0 \text{ m}$$

Hence $\Delta y = \lambda \cdot \dfrac{D}{d}$

(iii) From the above relationship, if f doubles, the wavelength of the waves (λ) halves and so Δy halves also $\Rightarrow \Delta y = 1.0$ m.
(iv) From the above relationship, if d doubles, the value of Δy halves (inverse proportion) $\Rightarrow \Delta y = 1.0$ m.

W4
(a) The telescope has to be adjusted to receive parallel light. This is done by taking it to an open window of the laboratory and focusing it upon a distant object (but not the sun!). The telescope is then used to observe the illuminated slit of the colli-

mator. The collimator is adjusted to give a focused image when looking through the telescope, and the slit is adjusted to give a suitable width. The collimator is now *producing* a parallel beam of light because the telescope has been set to receive parallel light.
(b) A bright (zero order) image of the slit is seen – a bright yellow light only.
(c) The first order line spectrum consists of a number of spectral lines representing the different wavelengths of the light emitted by the source. Since λ and $\sin \theta$ are directly proportional to each other (see d), there will be an angular separation of the red to blue lines, with the blue lines at the smaller angle because of the smaller λ.
(d) n refers to the order of the spectrum : $n = 1, 2$, etc. ;
λ refers to the wavelength of the light from the source ;
d refers to the spacing between the slits of the diffraction grating.
(e) In position A, a bright white line will be seen. At the position of the first order spectrum, a **continuous spectrum** for white light will be found, with blue nearer to A as before.

Practice Questions

1. (W1)
Microwaves travelling at 3.0×10^8 m s^{-1} in air and having a wavelength of 3.0 cm are directed towards a tank filled with light oil. The waves meet the tank at an incident angle of 45° and undergo refraction such that the angle of refraction is 30°.
(a) Calculate (i) the frequency of the microwaves in the oil, (ii) the speed and wavelength of the microwaves in the oil.
(b) Why would this experiment fail if the tank had metal sides?

2. (W2)
(a) What is the condition for a maximum on the screen used in a Young's double-slit interference experiment?
(b) In the relationship $\Delta y = \lambda . \dfrac{D}{d}$, what does

(i) Δy represent? (ii) λ represent? (iii) D represent? (iv) d represent?
(c) Suggest typical values for λ, D and d in an experiment carried out in class.
(d) What is meant by the phrase 'monochromatic light source'?

3. (W2)
A helium-neon laser (a source of intense monochromatic light) is used to produce an interference pattern in a Young's double-slit experiment. From the information supplied below, calculate the wavelength of the light from the laser.

Distance from screen to double slit = 4.00 m
Width of 10 fringes = 0.100 m
Separation of the slits = 2.5×10^{-4} m

4. (W4)
The diffraction grating in problem W4 is replaced by a glass prism.
(a) What main differences would this make to what is seen when the telescope is moved slowly around the table of the spectrometer?
(b) Why does a prism produce a line spectrum of sodium light and a continuous spectrum of sunlight?

5. (W4)
A diffraction grating having a spacing of 3.3×10^{-6} m is set up on a spectrometer as in problem W4. A filter is used to transmit a strong blue line from a cadmium source. The two first order maxima are separated by an angle of 17.2°.
(a) Find the wavelength of the blue line in cadmium.
(b) Determine the angle between the two second order maxima.

UNIT M—Mechanics

Worked Examples

M1 The stroboscope

(a) A boy ties a white ball to a piece of string and then whirls it at a steady rate in a vertical circle. When light from a stroboscope set at 12 flashes per second is directed towards the ball, *four* apparently stationary images of the ball are seen. It is also tested that no higher flash-rate can produce this pattern.

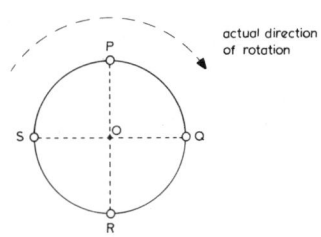

    (i) At what rate is the ball revolving about O? (2)
    (ii) If the length of the string is 0.50 m, what is the speed of the ball in m s^{-1}? (1)
    (iii) The boy claims that the velocity of the ball at P is different from its velocity at R. Is he right or not? (2)

(b) Mr. Flashman took the following stroboscope photograph of a simple pendulum while it was making one complete oscillation, to and fro. The setting of the stroboscope was 10 f.p.s.

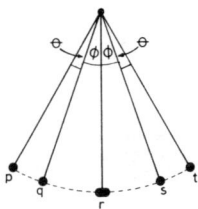

    (i) Calculate the period and the frequency of this oscillation. (2)
    (ii) Explain why the angles labelled ϕ are larger than those labelled θ. (2)
    (iii) Why is the image slightly distorted at r, and yet spherical at both p and t? (1)

M2 Velocity-time graphs

The following two graphs show how the velocities of two cars vary with time. Both cars start and stop at the same time, and move along a straight road.

Worked examples

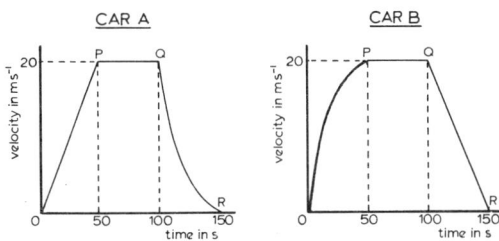

(a) Describe in general terms the motion of Car A over the three sections OP, PQ and QR of its graph. (2)
(b) Do the same for Car B. (2)
(c) Which car travelled farther? Explain your answer. (1)
(d) Calculate the acceleration, in m s^{-2}, of
   (i) Car A during part OP of its motion, and (ii) Car B during part QR. (2)
(e) If the average velocity of Car A is known to be 12 m s^{-1} for its entire journey, what distance does it travel during part QR? (2)
(f) If the distance travelled by Car B during part OP of its journey is 600 m, calculate its average velocity for its entire journey. (1)

M3 Uniform acceleration

Bob Slade is hurtling down a straight section of a Cresta Run which has two tunnels, as shown. He enters TUNNEL I at 15 m s^{-1} and emerges from it 4 seconds later.

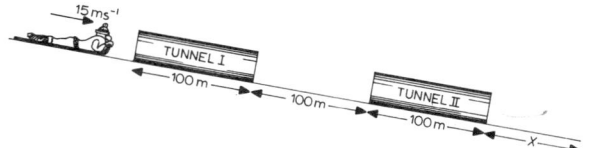

Assuming that Bob is accelerating uniformly, calculate
(a) the speed at which he (i) enters TUNNEL II and (ii) leaves TUNNEL II. (5)
(b) the time he spends moving through TUNNEL II. (3)
(c) the distance x to the foot of the slope if he arrives 1.0 s after emerging from TUNNEL II. (2)

M4 The effects of friction on slopes

A trolley of mass 1.5 kg is released from rest and it rolls down a board which is angled at 30° to the horizontal.
The following table gives some information about the trolley's journey.

Time from start (in s)	0.2	0.4	0.6	0.8
Distance gone from rest (in m)	0.08	0.32	0.72	1.28

(a) Calculate the acceleration of the trolley in m s^{-2}. (3)
(b) What would be the acceleration of this trolley if friction were absent? (2)
(c) Calculate the value of the frictional force which is acting upon the trolley. (2)
(d) What would you do to friction-compensate the trolley? (3)

M5 Motion in a lift

A hotel hoist which is cable-operated has a fully-loaded mass of 200 kg. For safety reasons, the tension in the cable must not exceed 3000 N.

(a) What is the tension in the cable, in newtons, when the hoist is
 (i) at rest? (2)
 (ii) moving upwards at a uniform speed? (1)
 (iii) moving downwards at a uniform speed? (1)
 (iv) accelerating upwards at 2.0 m s^{-2}? (2)
 (v) accelerating downwards at 2.0 m s^{-2}? (2)
(b) Calculate the maximum permitted acceleration while the hoist is moving up from the kitchen to the dining room. (2)

M6 Towing – the forces involved

Mr Fortune wins a dazzling new sports car in a newspaper competition and he celebrates by going on a caravan holiday using his own, not so dazzling, caravan. The mass of the car when occupied is 1000 kg and the mass of the caravan is 800 kg.

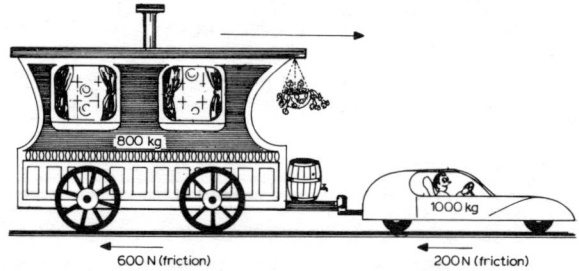

Holiday situation A : Mr Fortune accelerates jauntily away from green traffic lights at 2.0 m s^{-2}.
Holiday situation B : The car crawls along steadily at 4.0 m s^{-1} in a huge traffic snarl-up.
Holiday situation C : The car decelerates at 6.0 m s^{-2} when Mr Fortune brakes to avoid a runaway steamroller.
The frictional forces on the car and the caravan are 200 N and 600 N respectively.
(a) Estimate the forward thrust of the car engine for
 (i) situation A and (ii) situation B. (5)
(b) Estimate the horizontal force exerted on the tow socket of the caravan in each situation. (5)

M7 Projectile motion involving graphs

A projectile is fired across level country and it takes 6 s to travel from x to z. The highest point reached is labelled y. The air resistance on this projectile is negligible.

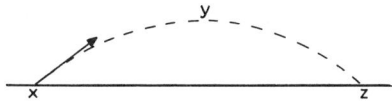

Graphs of the horizontal and vertical components of velocity during its flight are shown for the projectile:

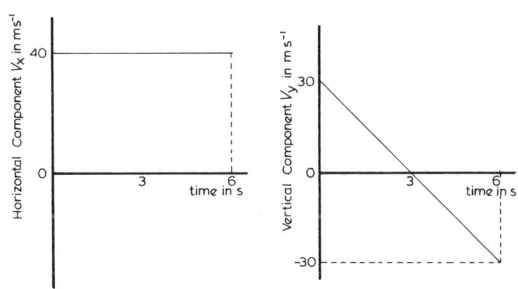

(a) What does the first graph tell you about the horizontal motion of the projectile? (1)
(b) What does the second graph tell you about the vertical motion of the projectile? (1)
(c) Using a vector diagram, find the speed and angle at which the projectile was fired from point x. (3)
(d) Explain why the speed of the projectile is least at position y, and find a value for this speed. (2)
(e) How high above the ground is point y? (2)
(f) What is the range xz of the projectile? (1)

M8 More about projectiles

Fred climbs up to the top of a cliff and fires a pebble straight out from its edge. It hits the beach at an angle of 30°.

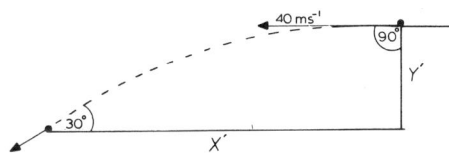

(a) What is the speed of the pebble just before impact with the beach? (3)
(b) Calculate the time of flight. (3)
(c) What is the height (Y') of the cliff? (2)
(d) How far out from the base of the cliff (X') does the pebble land? (2)

M9 Momentum – the inelastic collision

A trolley of mass 2.0 kg, travelling to the right at 4.0 m s^{-1}, collides head-on with a trolley of mass 0.5 kg travelling in the opposite direction. The trolleys have the same kinetic energy before impact, and they move off as a single unit after the impact (inelastic collision).

20 Mechanics

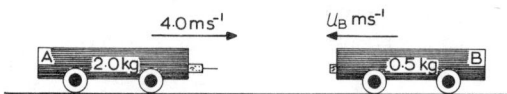

(a) Calculate the initial approach velocity of the lighter trolley. (3)
(b) What is the common velocity of the trolleys after the collision? (3)
(c) Calculate the loss of kinetic energy which occurs during the collision and try to account for this loss. (4)

M10 Introducing impulse

The **impulse** of a force is defined as the product of the **force** and the **time** for which it acts.
(a) Suggest suitable SI units for **impulse**. (1)
(b) During a sponsored piano race for charity a piano of mass 500 kg, initially at rest, is accelerated by a resultant (unbalanced) force of 900 N for a time of 5.0 s.

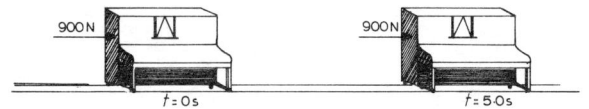

(i) Show that the change in momentum of the piano is equal to the applied impulse. (3)
(ii) How much work is done by the resultant force in 5.0 s? (3)
(iii) Calculate the kinetic energy of the piano after 5.0 s. (3)

M11 The linear air track

(a) A linear air track vehicle is shown travelling towards a solid barrier at a constant speed of 0.6 m s^{-1}. In position A, it is 1.2 m from the barrier. After hitting the barrier, the vehicle rebounds at 0.4 m s^{-1}.

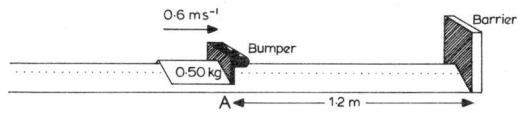

The time taken for the vehicle to travel from A up to the barrier, and then back to A is 5.5 s.
(i) Calculate the change in momentum of the vehicle due to the impact. (2)
(ii) What force, in newtons, is exerted by the wall on the vehicle? (4)
(b) The same vehicle, travelling at the same speed as before, collides with a second, but stationary, vehicle of mass 0.30 kg. Its speed is reduced to 0.15 m s^{-1}.

(i) Calculate the speed of the 0.30 kg vehicle after the collision. (2)
(ii) What is the impulse, in N s, applied to the 0.30 kg vehicle during impact? (2)

M12 Two-dimensional collisions

During an experiment on 2-dimensional collisions, two identical pucks, A and B, collide with each other on a frictionless air-table. The holes in the air-table provide a convenient reference grid of 1 cm squares. Puck A, whose speed and direction are not shown, collides with the stationary puck B at point P on the table. Exactly one second after the impact, the pucks are in the positions A' and B', as shown.

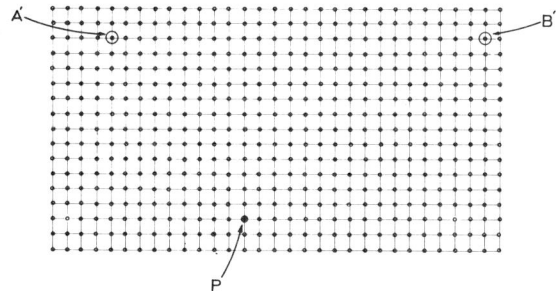

(a) Using the grid for any necessary measurements, calculate the speed and direction of puck A *before* the impact occurs. (7)
(b) Compare the kinetic energies of the pucks before and after the impact, and comment upon this (3)

M13 The bouncing ball

The following velocity-time graph refers to the motion of a soft rubber ball of mass 0.2 kg which was dropped on to a level pavement.

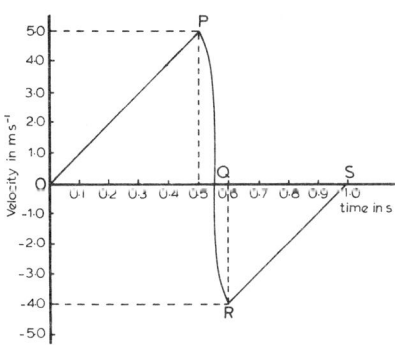

(a) Describe in detail the motion of the ball, referring to sections OP, PQ, QR and RS for convenience. (3)
(b) Calculate the initial height of the ball above the pavement. (2)
(c) What height does the ball reach after the first bounce? (2)
(d) How much mechanical energy is lost by the ball during its impact with the pavement? (1)
(e) Show that the deceleration of the ball while it is being brought to rest by the pavement is 100 m s^{-2}. (2)

M14 Energy transformations
Bob Slade starts his final run on the frictionless speed course by pushing his sledge off from the start A at 5.0 m s^{-1}. For the duration of the run ABCD, he surprises himself by actually remaining on the track. Eventually he crosses the finishing-line at D and then brakes on the horizontal straight.

22 Mechanics

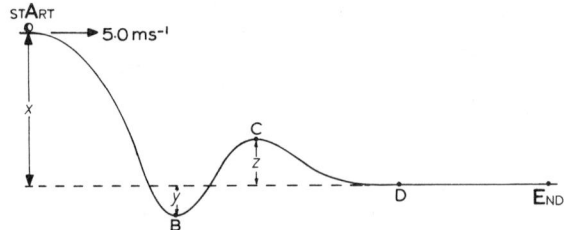

(a) If the vertical distances x, y and z are 50 m, 10 m and 15 m respectively, find Bob's speed at (i) B, (ii) C and (iii) D. (6)

(b) The total mass of Bob and his sledge is 80 kg. What constant braking force applied over part DE, a distance of 50 m, will stop the sledge at E? (4)

M15 Work and power

Penelope Farthing decides to adapt her bicycle so that she can measure her useful power output. She fits a belt over a smooth-edged wheel and connects each end to a newton-balance, as shown.

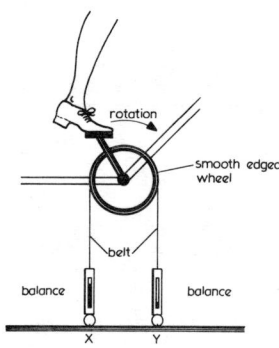

While Penelope pedals at a steady rate, her cycling friend Isabel Wringing records the following information :

Circumference of wheel = 0.80 m
Reading on balance X = 80 N
Reading on balance Y = 20 N
Number of revolutions pedalled = 270
Time for trial = 3 minutes

(a) What is the value of the friction force acting between the belt and the wheel? (2)
(b) How much work does Penelope do against the belt during this trial? (3)
(c) What is her useful power output against the belt? (2)
(d) Explain what happens to the energy which she supplies to overcome the frictional force. (3)

M16 Power output of an engine

(a) During a tour of Europe, a loaded bus of total mass 24 000 kg crawls at a steady speed up an Alpine road which is signposted as '1 in 8'. This means that the bus is rising by 1 m for every 8 m driven along the road.

The total frictional resistance acting upon the bus is 15 000 N. Estimate the power output of the engine if the speed of the bus is 2.0 m s^{-1}. (7)

(b) During the return journey, the bus has to go down the same mountain road, but unfortunately traffic is very slow due to some roadworks farther down. The driver lets the bus free-wheel steadily at 1.0 m s⁻¹ by applying the brakes gently. Estimate the new total frictional resistance acting upon the bus. (3)

M17 Calculating gas pressure

Elspeth is asked by her physics teacher to measure some gas pressures using the equipment shown.

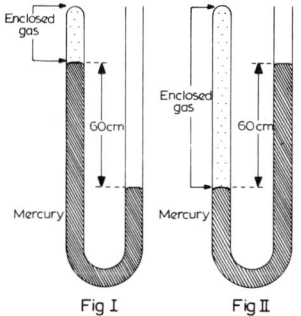
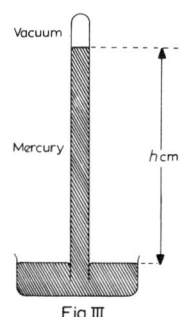

Fig I   Fig II   Fig III

The atmospheric pressure, p_0 = 1.0 x 10⁵ Pa (N m⁻²)
The density of mercury, ρ = 1.36 x 10⁴ kg m⁻³
The acceleration due to gravity g = 10 m s⁻²

(a) What value should Elspeth get for the pressure of the enclosed gas in the apparatus shown in Fig. I? (2)
(b) What value should she get for the pressure of the enclosed gas in the situation shown in Fig. II? (2)
(c) Elspeth then attempts to *calculate*, rather than measure, the height of the mercury column (h) for the simple barometer shown in Fig. III. Which method should she use? (3)
(d) Next, the apparatus shown in Fig. III is completely enclosed in a bell jar from which air is extracted by a pump. The mercury column drops until h is only 5 mm high. What pressure exists inside the bell jar at this stage? (3)

Solutions to Worked Examples

M1

(a) (i) The flash-rate is 12 f.p.s. and so the ball takes $\frac{1}{12}$ s to move from P to Q. An entire revolution therefore takes four times as long, i.e. $\frac{1}{3}$ s. Therefore the ball makes 3 revolutions per second.

(ii) speed = $\dfrac{\text{distance}}{\text{time}}$ = $\dfrac{\text{circumference}}{\text{time per rev.}}$ = $\dfrac{2 \times \pi \times 0.50}{\frac{1}{3}}$ = 3π m s⁻¹

(iii) Yes, he is right because **velocity** is a **vector** and refers to speed in a particular direction. The ball has the same **speed** at P and R, but different velocities because it is moving to the right at P, and to the left at R.

(b) (i) The flash-rate is 10 f.p.s. ⇒ time between images = 0.1 s.
time for bob to move from p to t = 4 x 0.1 = 0.4 s
time for entire oscillation, to and fro, = 0.8 s (the period)

24   *Mechanics*

frequency $= \dfrac{1}{\text{period}} = \dfrac{1}{0.8} = 1.25$ Hz

(ii) The bob is moving faster over the central part of its swing than when it is near the ends of its swing ∴ it covers a greater distance during time interval q r than it does during the same time interval p q.
(iii) The actual flash takes a short time and during this time the bob is moving at r. This results in a slightly elongated image. At p and t, the ball is moving so slowly during the flash that the effect is not noticeable.

M2
(a) *Car A* accelerates uniformly during OP, moves with a constant velocity during PQ and moves with a decreasing deceleration during QR until it stops.
(b) *Car B* moves with decreasing acceleration during OP, then travels with constant velocity during PQ and decelerates uniformly to rest during QR.
(c) *Car B* travels farther since the area under its *v-t* graph is greater.
(d) (i) *Car A* : $a = \dfrac{(v-u)}{t} = \dfrac{(20-0)\text{ m s}^{-1}}{50\text{ s}} = 0.4$ m s^{-2}

   (ii) *Car B* : $a = \dfrac{(v-u)}{t} = \dfrac{(0-20)\text{ m s}^{-1}}{50\text{ s}} = -0.4$ m s^{-2}

(e) total displacement = average velocity x time = 12 m s^{-1} x 150 s = 1800 m
∴ 1800 = ($\tfrac{1}{2}$ x 50 x 20) + (50 x 20) + X, where X is area under QR.
X = 300 m ∴ distance gone during QR is 300 m.
(f) total displacement = total area under graph = 600 + 1000 + 500 = 2100 m

total time taken = 150 s ⇒ average velocity of *Car B* = $\dfrac{2100 \text{ m}}{150 \text{ s}}$ = 14 m s^{-1}

M3
(a) Consider the motion through TUNNEL I. Since the acceleration is uniform.
$s = ut + \tfrac{1}{2}at^2$ ⇒ $100 = (15 \times 4) + \tfrac{1}{2} \times a \times (4)^2$ ⇒ $a = 5$ m s^{-2}
Now that we know the acceleration, we can find the speeds for TUNNEL II.
  (i) *Entering TUNNEL II'* $v^2 = u^2 + 2as = (15)^2 + (2 \times 5 \times 200)$ ⇒ $v = 47.2$ m s^{-1}
  (ii) *Leaving TUNNEL II* $v^2 = u^2 + 2as = (15)^2 + (2 \times 5 \times 300)$ ⇒ $v = 56.8$ m s^{-1}
(b) We now know that he enters this tunnel at 47.2 m s^{-1}, and leaves at 56.8 m s^{-1}. We also know that the acceleration is 5 m s^{-2}. We find the time as follows :

$t = \dfrac{(v-u)}{a} = \dfrac{(56.8-47.2)\text{m s}^{-1}}{5 \text{ m s}^{-2}} = 1.9$ s

(c) He starts covering distance x at 56.8 m s^{-1} and takes 1.0 s to reach the foot.
$x = ut + \tfrac{1}{2}at^2 = (56.8 \times 1.0) + \tfrac{1}{2} \times 5 \times (1.0)^2 = 59.3$ m.

M4
(a) *Method I*

Average velocity over the first 0.08 m is $v_1 = \dfrac{d_1}{t} = \dfrac{0.08 \text{ m}}{0.2 \text{ s}} = 0.4$ m s^{-1}

Average velocity over the next 0.24 m is $v_2 = \dfrac{d_2}{t} = \dfrac{0.24 \text{ m}}{0.2 \text{ s}} = 1.2$ m s^{-1}

acceleration = $\dfrac{\text{change in velocity}}{\text{time for change}} = \dfrac{(1.2-0.4) \text{ m s}^{-1}}{0.2 \text{ s}} = 4.0$ m s^{-2}

Method II
By inspecting the pattern of distances gone, we see that the acceleration is uniform so we can use the equations of motion.
$s = ut + \frac{1}{2}at^2 = \frac{1}{2}at^2$, since $u = 0$ in this example

$$a = \frac{2s}{t^2} = \frac{2 \times 1.28}{(0.8)^2} = 4.0 \text{ m s}^{-2} \text{ (using } s = 1.28 \text{ m and } t = 0.8 \text{ s)}$$

(b) In the absence of friction the acceleration down a slope is $a = g \sin \theta$, since the component of the trolley's weight acting down the slope is $mg \sin \theta$. From Newton's 2nd Law,

$$a = \frac{F}{m} = \frac{mg \sin \theta}{m} = g \sin \theta$$

When the slope is angled at $30°$, $a = 10 \times \sin 30 = 5 \text{ m s}^{-2}$.

(c) resultant force on trolley = (weight component − frictional force)
$\Rightarrow F = (mg \sin \theta - P) = ma$
$1.5 \times 10 \times \sin 30 - P = 1.5 \times 4.0 \Rightarrow P = 7.5 - 6.0 = 1.5 \text{ N}$

The frictional force on the trolley is 1.5 N.

(d) To compensate for friction, we must tilt the board to a special angle θ_{fc} so that the frictional force P is just balanced by the weight component down the slope. Thus

$$\boxed{mg \sin \theta_{fc} = P}$$

M5
(a) (i) When the hoist is at rest, the upwards tension T in the cable must just support the downwards weight W of the hoist.

$T = W = mg = 200 \times 10 = 2000 \text{ N}$

(ii) & (iii) In both cases, the hoist is moving at uniform velocity and so it is **not accelerating**. According to Newton's Laws, this means that the hoist has **no resultant force** acting on it. The tension T must, therefore, just balance out the weight as in (i).

Resultant force,
$F = (T - W) = 0 \Rightarrow T = W = 2000 \text{ N}$

(iv) The hoist is accelerating upwards and so the resultant force F must be acting upwards. Obviously, T has to be greater than W for this to happen.
Applying Newton's 2nd Law,
$F = (T - W) = ma = 200 \times 2.0 = 400 \text{ N}$

The tension,
$T = 400 + W = 400 + 2000 = 2400 \text{ N}$

(v) The hoist is accelerating downwards and so the resultant force F must be acting downwards. Obviously, T has to be less than W for this to happen.

Resultant force,
$$F = (W - T) = ma = 200 \times 2.0 = 400 \text{ N}$$
The tension,
$$T = W - 400 = 2000 - 400 = 1600 \text{ N}$$
Force diagrams :

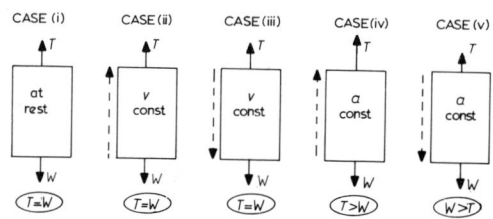

(b) The maximum value of acceleration, a_{max}, occurs when $T = 3000$ N. Comparing with (a) (iv), we see that
$$F = (T - W) = ma_{max}$$
$$a_{max} = \frac{(3000 - 2000)}{200} = 5.0 \text{ m s}^{-2}$$

M6

(a) (i) The entire mass of the system is 1800 kg and it accelerates at 2.0 m s^{-2}. Thus the resultant force,
$$F = m_{TOTAL} \times a = 1800 \times 2.0 = 3600 \text{ N}$$
resultant force (F) = forward thrust of engine (Q) – entire friction force (P)
$$F = Q - P \Rightarrow 3600 = Q - (600 + 200) \Rightarrow Q = 4400 \text{ N}$$
The thrust of the engines is 4400 N in this situation.
(ii) The system is moving at a **steady velocity** and so there is **zero resultant force** to cause acceleration. Since $F = 0$, $Q = P = 800$ N. The thrust overcomes the frictional forces and is therefore 800 N.

(b) Consider the horizontal force acting upon the caravan. There will be a forward pull X acting through the tow bar, and an opposing force P due to the friction on the caravan.

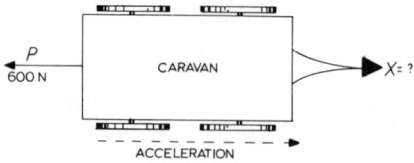

Situation A Here $F = (X - P) = ma$ ($m = 800$ kg for caravan).
Substituting,
$$X - 600 = 800 \times 2.0 \Rightarrow X = 2200 \text{ N}$$

Situation B Here $X = P$ because the caravan is not accelerating. Thus the required force $X = 600$ N.

Situation C The caravan is decelerating and so the resultant force F must be acting backwards (to left). Its value is $F = ma = 800 \times 6.0 = 4800$ N. The frictional force on the caravan supplies 600 N of this so the additional 4200 N must be transmitted through the tow bar $\Rightarrow X = 4200$ N (to left).

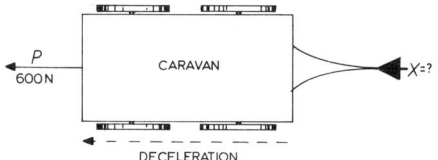

M7
(a) The horizontal velocity is constant throughout the motion. This is because there are no horizontal forces acting upon the projectile while it is in flight (air resistance is negligible). $v_x = 40$ m s^{-1}.
(b) The vertical velocity is steadily reduced to zero during the upwards part of the flight, and then steadily increased again during the downwards part of the flight. The acceleration and deceleration are of value 10 m s^{-2} (the acceleration due to gravity).
(c) The initial values of v_x and v_y are 40 m s^{-1} and 30 m s^{-1}.
Velocity vector diagram :

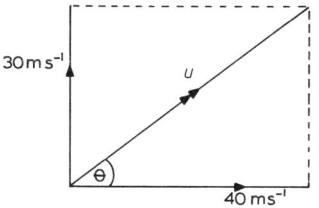

The initial speed is given by $u = \sqrt{40^2 + 30^2} = 50$ m s^{-1}.
The initial angle of projection is given by θ, where
$$\tan \theta = \frac{30}{40} = 0.75 \Rightarrow \theta = 37°$$

(d) The total mechanical energy (k.e. plus p.e.) stays constant. At y, it has its greatest amount of p.e. $\therefore$ it has its least amount of k.e. Thus its speed is least at y.
At y, the projectile has horizontal motion only $\Rightarrow v = v_x = 40$ m s^{-1}.
(e) The projectile takes 3 s to drop from y to z. It would take the same time (3 s) if it were simply dropped straight down from y.
height $= h = \tfrac{1}{2}gt^2 = \tfrac{1}{2} \times 10 \times (3)^2 = 45$ m
(f) Since the horizontal velocity is always 40 m s^{-1}, the projectile must be moving horizontally at 40 m s^{-1} for 6 s. It therefore travels a distance of 240 m $\Rightarrow$ $xz = 240$ m.

M8
(a) Just before impact, it is travelling at 30°, and its horizontal component of velocity is 40 m s^{-1}. The vector diagram is

28 Mechanics

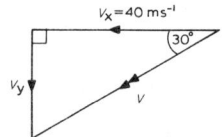

Obviously, $v = \dfrac{40}{\cos 30} = 46.2$ m s^{-1} and $v_y = 40 \tan 30 = 23.1$ m s^{-1}.

Pebble is travelling at 46.2 m s^{-1} just before it hits the beach.
(b) initial vertical velocity $= u_y = 0$ m s^{-1}, as pebble starts horizontally.
final vertical velocity $= v_y = 23.1$ m s^{-1}, from (a).
Applying $v = u + at$ to the vertical motion, we get
$23.1 = 0 + 10t \Rightarrow t = 2.31$ s

(c) To find the height, we apply $v^2 = u^2 + 2as$ to the vertical motion :
$(23.1)^2 = (0)^2 + 2 \times 10 \times Y' \Rightarrow Y' = 26.7$ m

(d) Pebble travels along at 40 m s^{-1} for 2.31 s $\Rightarrow X' = 40 \times 2.31 = 92.4$ m

M9

(a) kinetic energy of trolley A before impact $= \tfrac{1}{2} m_A u_A^2 = \tfrac{1}{2} \times 2.0 \times (4.0)^2 = 16.0$ J
Thus k.e. of trolley B before impact $= 16.0$ J, also.
$\Rightarrow \tfrac{1}{2} m_B u_B^2 = \tfrac{1}{2} \times 0.5 \times u_B^2 = 16.0 \Rightarrow u_B = 8.0$ m s^{-1}

The smaller trolley approaches with a velocity of 8.0 m s^{-1} to the left.
(b) total momentum before impact $= m_A u_A - m_B u_B = (2.0 \times 4.0) - (0.5 \times 8.0)$
$= 8.0 - 4.0 = 4.0$ kg m s^{-1}, to the right.
(Remember that momentum is a **vector** quantity, and so the direction must be taken into account. Here we use right is + ve and left is − ve).
total momentum after impact $= (m_A + m_B) \times V = 2.5 V$ kg m s^{-1}
Here V represents the common velocity after impact.
By the Principle of Conservation of Momentum,
$2.5 V = 4.0 \Rightarrow V = 1.6$ m s^{-1}, to the right.

(c) Although total (vector) momentum is always conserved in collisions, kinetic energy is only conserved when the collision is perfectly elastic, and this does not happen very often. Using k.e. $= \tfrac{1}{2} mv^2$ and remembering that it is a scalar quantity,
total k.e. of system before impact $= 16$ J $+ 16$ J $= 32$ J (see (a))
total k.e. of system after impact $= \tfrac{1}{2} \times (2.0 + 0.5) \times 1.6^2 = 3.2$ J
loss in k.e. during impact $= (32 - 3.2)$ J $= 28.8$ J
This loss of k.e. is due to the production of heat energy, sound energy, deformation energy, etc.

M10

(a) **impulse = force × time**. The SI units for force and time are newtons and seconds respectively, and so impulse has the unit newtons × seconds, usually written as simply N s.

(b)
(i) acceleration of piano $= a = \dfrac{F}{m} = \dfrac{900 \text{ N}}{500 \text{ kg}} = 1.8$ m s^{-2}

The velocity gained by the piano in 5.0 s is obtained from

$v = u + at = 0 + 1.8 \times 5.0 = 9.0$ m s^{-1}
final momentum = mass x final velocity = mv = 500 x 9.0 = 4500 kg m s^{-1}
applied impulse = force x time = 900 x 5.0 = 4500 N s
Thus we have shown that **change in momentum produced = applied impulse**
(*Note* : It follows from this that **force = rate of change of momentum**)
(ii) work done by resultant force = resultant force x (distance moved in 5.0 s)
$E = Fs = 900 s$
But $s = ut + \frac{1}{2}at^2 = 0 + \frac{1}{2} \times 1.8 \times (5.0)^2 = 22.5$ m.
Therefore $E = 900 \times 22.5 = 20\,250$ J.
The resultant force does 20 250 J of work on the piano.
(iii) final k.e. = $\frac{1}{2} mv^2 = \frac{1}{2} \times 500 \times (9.0)^2 = 20\,250$ J, which is the expected answer for the k.e. of the piano, because the work done on the piano by the resultant force is transformed into an equal amount of energy of motion.

M11
(a) (i) initial momentum = mu = 0.50 x 0.6 = 0.3 kg m s^{-1} to right.
final momentum = mv = 0.50 x (– 0.4) = – 0.2 kg m s^{-1} to left.
change in momentum = $\Delta mv = (mv - mu) = -0.2 - 0.3 = -0.5$ kg m s^{-1}
(The minus sign means in effect that the force acted to the left on the vehicle).

(ii) time taken for vehicle to get to barrier from A = $\dfrac{1.2 \text{ m}}{0.6 \text{ m s}^{-1}}$ = 2.0 s

time for vehicle to bounce back to A = $\dfrac{1.2 \text{ m}}{0.4 \text{ m s}^{-1}}$ = 3.0 s

time for impact = 5.5 – 2.0 – 3.0 = 0.5 s.
Impulse applied to vehicle = change in momentum = – 0.5 kg m s^{-1} (N s).
force x time = $F \times 0.5$ s = – 0.5 N s $\Rightarrow F$ = – 1.0 N.
(Again the minus sign indicates that the force acts to left on vehicle).
(b) (i) total momentum before impact = total momentum after impact.
$0.50 \times 0.6 = (0.50 \times 0.15) + (0.30 \times V) \Rightarrow V = 0.75$ m s^{-1}
(ii) impulse = change in momentum = 0.30 x 0.75 = 0.225 N s.

M12
(a) Measurements from the grid are as follows :

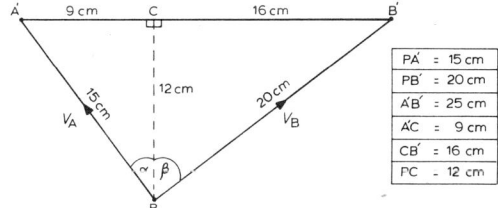

v_A =	$\dfrac{15 \text{ cm}}{1 \text{ s}}$ = 15 cm s^{-1}	$\tan \alpha$ =	$\dfrac{9 \text{ cm}}{12 \text{ cm}}$ = 0.75 $\Rightarrow \alpha$ = 36.9°
v_B =	$\dfrac{20 \text{ cm}}{1 \text{ s}}$ = 20 cm s^{-1}	$\tan \beta$ =	$\dfrac{16 \text{ cm}}{12 \text{ cm}} = \dfrac{4}{3} \Rightarrow \beta$ = 53.1°

The speed and direction of A before impact can be found either by calculation using the conservation of momentum principle, or by a vector drawing. In the following diagram, we use **velocities** rather than **momenta**. This is allowed because the pucks have identical masses.
From measurement, we obtain $u_A = 25$ cm s^{-1}, and $\theta = 73.7°$.

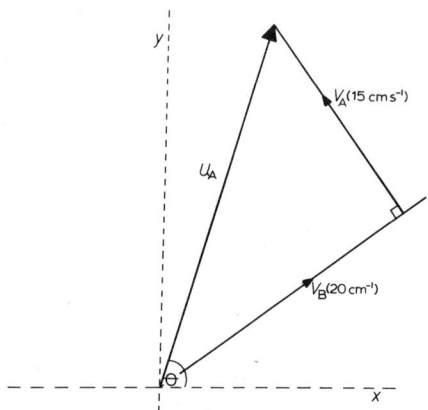

(b) Let the mass of each puck be m.
k.e. before $= \frac{1}{2} \times m \times (25)^2 = 312.5 \, m$ units
k.e. after $= \frac{1}{2} \times m \times (15)^2 + \frac{1}{2} \times m \times (20)^2 = 312.5 \, m$ units
Thus k.e. before = k.e. after, so the collision is perfectly elastic. This is more immediately obvious from the fact that the triangle of velocities is right-angled and $u_A^2 = v_A^2 + v_B^2$

M13
(a) *OP* : The ball is released and accelerates uniformly due to gravity for 0.5 s, reaching a maximum *downwards* velocity of 5.0 m s^{-1}.
PQ : The ball is in contact with the pavement and is being brought to rest in 0.05 s.
QR : The ball is still in contact with the pavement and is now being accelerated rapidly *upwards*, reaching 4.0 m s^{-1} in 0.05 s.
RS : The ball leaves the pavement at 4.0 m s^{-1} and travels upwards, being decelerated uniformly by gravity for 0.4 s until it reaches its new highest point.

(b) Applying the equation of motion $s = ut + \frac{1}{2}at^2$ to part OP, we get
$s = (0 \times 0.5) + \frac{1}{2} \times 10 \times (0.5)^2 = 1.25$ m. This is initial height.
(Alternatively, calculate the area under OP of this v–t graph :
area under OP $= \frac{1}{2} \times 0.5 \times 5.0 = 1.25$ m).

(c) This time we consider part RS :
$s = ut + \frac{1}{2}at^2 = (4.0 \times 0.4) - \frac{1}{2} \times 10 \times (0.4)^2$
$s = 0.8$ m, which is the new height. (Again the area under RS gives this).

(d) loss of k.e. upon impact $= \frac{1}{2} \times 0.2 \times (5.0)^2 - \frac{1}{2} \times 0.2 \times (4.0)^2 = 0.9$ J.

(e) The velocity of the ball is reduced from 5.0 m s^{-1} to 0 m s^{-1} in PQ.
From graph, the time for this reduction is 0.05 s.

$$\text{deceleration} = \frac{\text{loss in velocity}}{\text{time for loss}} = \frac{5.0 \text{ m s}^{-1}}{0.05 \text{ s}} = 100 \text{ m s}^{-2}.$$

M14

(a) k.e. at A = $\frac{1}{2} \times m \times (5.0)^2$ = 12.5m joules.

(i) On his trip from A to B, Bob and his sledge lose potential energy and so gain more kinetic energy,
k.e. at B = k.e. at A + p.e. lost
$\frac{1}{2} mv_B^2$ = 12.5m + $mg(x + y)$ = 12.5m + $m \times 10 \times (50 + 10)$
$\frac{1}{2} mv_B^2$ = 12.5m + 600m ⇒ v_B^2 = 612.5 × 2 ⇒ v_B = 35 m s^{-1}

(ii) As in (i), but p.e. lost is $mg(x - z)$ = $mg \times (50 - 15)$.
This gives v_C = 26.9 m s^{-1}.

(iii) As in (i), but p.e. lost is mgx = $mg \times 50$. This gives v_D = 32.0 m s^{-1}.

(b) energy possessed at D = 12.5 m + mgx = 512.5m joules
= 512.5 × 80 = 41 000 J
work done in stopping sledge = braking force × braking distance
= $F \times S$ = $F \times 50$ = 50F joules
Thus, we see that 50F = 41 000 ⇒ F = 820 N.

M15

(a) friction force = difference in tensions = (80 − 20) N = 60 N.
(b) work done per revolution = force × circumference = 60 N × 0.80 m = 48 J.
total work = work per rev. × no. of revs. = 48 J × 270 revs. = 12 960 J.
(c) useful power output = $\dfrac{\text{total work}}{\text{time}}$ = $\dfrac{12\,960 \text{ J}}{180 \text{ s}}$ = 72 W.
(d) The energy she supplies is converted into heat energy at the rim of the wheel by friction.

M16

(a) The engine has to overcome two forces acting along the slope against the direction of motion:

(i) the frictional resistance which is P = 15 000 N and
(ii) the component of the weight of the bus, Q, which acts down the slope.
By resolving the weight W into components Q and R, as shown, we see that $Q = W \sin \theta$ and $R = W \cos \theta$.

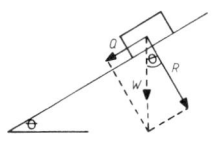

$$\sin \theta = \frac{Q}{W} \Rightarrow Q = W \sin \theta$$

$$\cos \theta = \frac{R}{W} \Rightarrow R = W \cos \theta$$

The entire force to be overcome by the engine is
$F = P + Q$ = 15 000 + Q = 15 000 + $mg \sin \theta$ = 15 000 + (24 000 × 10 × $\frac{1}{8}$)
F = 15 000 + 30 000 = 45 000 N

In *one second*, the bus moves a distance of 2.0 m against an entire force of 45 000 N.
Work done in one second = entire force × distance.
$E = Fs$ = 45 000 × 2.0 = 90 000 J

power developed = $\dfrac{\text{energy}}{\text{time}}$ = $\dfrac{90\,000 \text{ J}}{1 \text{ s}}$ = 90 000 W (or 90 kW)

(b) The bus is moving down the slope at a steady speed with the engine disengaged. The new frictional force must just be balancing the component of the weight acting down the slope, preventing acceleration.
New frictional force $= W \sin \theta = mg \sin \theta = 30\,000$ N.

M17
(a) The pressure of the atmosphere is supporting a 60 cm column of mercury *and* the pressure of the enclosed gas too.

atmospheric pressure = pressure of enclosed + pressure due to 60 cm
 gas   column of mercury

$\Rightarrow p_0 = p_{gas} + \rho gh \Rightarrow p_{gas} = p_0 - \rho gh$

$\Rightarrow p_{gas} = (1.0 \times 10^5) - (1.36 \times 10^4 \times 10 \times 0.60) = 1.84 \times 10^4$ Pa

(b) The pressure of the enclosed gas is supporting a 60 cm column of mercury *and* the pressure of the atmosphere too.

pressure of enclosed gas = atmospheric + pressure due to 60 cm
           pressure   column of mercury

$\Rightarrow p_{gas} = p_0 + \rho gh = (1.0 \times 10^5) + (1.36 \times 10^4 \times 10 \times 0.60)$

$\Rightarrow p_{gas} = 1.816 \times 10^5$ Pa

(c) There is a vacuum at the top of the enclosed tube. Thus the pressure of the atmosphere is supporting the pressure due to a column of mercury of height h (see Fig. III).

atmospheric pressure = pressure due to h metres of mercury

$p_0 = \rho gh \Rightarrow h = \dfrac{p_0}{\rho g} = \dfrac{1.0 \times 10^5}{1.36 \times 10^4 \times 10}$

Thus $h = 0.73$ m (or 73 cm). This predicted height is smaller than the average observed height of 76 cm. The difference is due mainly to
 (i) using $g = 10$ m s^{-2} instead of $g = 9.81$ m s^{-2} and
 (ii) using $p_0 = 1.0 \times 10^5$ Pa instead of $p_0 = 1.01325 \times 10^5$ Pa.

(d) Using the equation $p_{jar} = \rho gh = 1.36 \times 10^4 \times 9.81 \times 5 \times 10^{-3}$

$p_{jar} = 667$ Pa

Here we have used the more accurate figures quoted in (c).
With the approximate figures given at the start of the problem, it turns out that $p_{jar} = 680$ Pa.

Practice Questions

1. (M1)
A cooling fan in a car has five identical blades. The fan appears to be stationary when viewed by light from a stroboscope set at 25 flashes per second. When the stroboscope frequency is increased, the next 'stationary view' of the fan shows the fan with apparently 10 blades.
(a) At what rate is the fan actually rotating?
(b) What is the period of the fan's rotation?
(c) If one blade were to snap off, what would be the highest stroboscope frequency to produce four 'stationary' blades, assuming the same actual rotation rate as before?

2. (M1)
A ball falling freely is photographed using a camera and a stroboscope. If the actual distances between three consecutive images are 10.0 cm and 12.5 cm respectively find
(a) the stroboscope frequency and
(b) the next actual position of the ball.
(Assume that the acceleration due to gravity is 10 m s^{-2} in this question).

3. (M3)
Bob Slade pushes himself off at 2.0 m s^{-1} from the top of a downhill run on his homemade ski-jump. This straight run of 40 m increases Bob's speed to 22.0 m s^{-1}.
(a) What is his acceleration on this downhill run?
(b) What is his speed 30 metres from the start?
(c) How much time does Bob take to complete this downhill run?

4. (M3)
A ball bearing is dropped vertically and it enters a vertical cardboard tube which is 40 cm long. The ball is out of view for 0.2 s.
(a) What is the speed of the ball bearing as it enters the tube?
(b) What is its speed when it emerges from the tube?

5. (M5)
A man stands on a set of bathroom scales, which have been specially calibrated in newtons, in a lift in a large office block. Before the lift moves, he notices that the scales record 800 N. The lift then moves between 5 floors in one continuous journey.
(a) Near the beginning of the journey, the scales register 600 N.
Is the lift going up or down?
(b) At different stages of the journey, the scales register 600 N, 800 N and 1200 N. Calculate the man's accelerations for each of these three stages of the journey.
(c) Which of the balance readings given in (b) refers to the section when the man would be least aware of his motion in the lift?

6. (M6)
Mr Fortune tows his caravan up a hill at a steady speed of 4.0 m s^{-1}. The car has a mass of 1000 kg and the caravan has a mass of 800 kg. The frictional forces acting on the car and the caravan are 200 N and 600 N respectively. The total thrust of the car engine is 9800 N.
(a) What is the component of the weight of the car and caravan down the slope?
(b) Estimate the incline which the car is travelling up.

7. (M6)
A spacecraft of mass 5000 kg is hovering above the surface of a planet where the gravitational acceleration is 16 m s^{-2}.
(a) Calculate the engine thrust which is needed to accelerate the spacecraft vertically
  (i) upwards and (ii) downwards at 20 m s^{-2}.
(b) What horizontal thrust would be needed to make the hovering spacecraft accelerate horizontally at 16 m s^{-2}?

8. (M7)
In the table, the height of a projectile above the ground and its horizontal range are recorded for each of the six seconds of its flight.

Time (s)	0	1	2	3	4	5	6
Height above ground (m)	0	25	40	45	40	25	0
Horizontal range (m)	0	40	80	120	160	200	240

(a) What are the initial vertical and horizontal components of velocity for this projectile?
(b) Which of these components remains constant throughout the flight?
(c) What is the vertical component of the velocity after 3 s?

9. (M8)
A plane flying horizontally at 150 m s^{-1} drops a projectile when its horizontal displacement from the target is 1200 m. Assuming that the effects of air resistance can be ignored,
(a) at what height should the plane fly to hit the target?
(b) what is the projectile's velocity (speed and direction) when it hits the target?

10. (M9, 11)
A trolley of mass 2.0 kg fitted with a spring bumper collides with a wall and rebounds at 3.0 m s^{-1}. It is known that 16.0 J of kinetic energy are lost during the impact.
(a) What is the momentum of the trolley as it leaves the wall?
(b) Calculate the kinetic energy of the trolley before it collides with the wall?
(c) At what speed does the trolley approach the wall?
(d) If the actual collision lasts 0.2 s, what force is exerted by the wall on the trolley?

11. (M9)
Dick Deadeye fires a pellet from his high velocity rifle into a lump of plasticene of mass 1.000 kg which is fixed to a stationary air track vehicle of mass 0.500 kg. The loaded vehicle moves off at 0.1 m s^{-1} when the pellet embeds itself in the plasticene.
(a) Assuming that the mass of the pellet is 0.001 kg, calculate the speed at which the pellet entered the plasticene.
(b) How much kinetic energy does the entire system lose during the impact?
(c) What happens to this energy?

12. (M12)
Mr Fortune is rowing his wife across Loch Ness, heading west at a leisurely 1.0 m s^{-1} when his craft is hit by another boat propelled by his daughter Miss Fortune. She had been heading north at 4.2 m s^{-1}. The colliding boats lock together and move off as a single unit after the impact. The loaded mass of Mr Fortune's boat is 425 kg whereas the loaded mass of Miss Fortune's boat is only 375 kg.
(a) What type of collision is described here?
(b) Calculate the velocity (speed and direction) of the locked boats just after the collision. Draw a vector diagram.

13. (M12)
Three magnetic pucks are held together in a group, side by side on an air table. They each have a mass of 0.20 kg. When released they are repelled apart, two of them at 160° to each other with speeds of 0.6 m s^{-1} and 0.5 m s^{-1}.
(a) Find, by vector diagram, the velocity of the third puck.
(b) How can you explain the fact that the kinetic energy of the pucks is zero before release, but is quite substantial after release?

14. (M13)
A rubber ball of mass 0.20 kg is dropped from a height of 1.8 m on to a flat road surface. The ball rebounds to a height of 1.25 m after being in contact with the road for 0.1 seconds. Air resistance can be ignored in this problem.
(a) Draw a graph, including numerical values, of the velocity of the ball from the instant of release to the time when it has rebounded to 1.25 m.
(b) How much energy is lost by the ball on rebounding?

15. (M14)
Bob Slade modifies part C of the run described in problem M14 to see what the maximum height of z would be if he could just reach C and no more. The rest of the run is the same and the starting speed from A is still 5.0 m s^{-1}. What would the maximum height be for z?

16. (M15)
Penelope Farthing attempts to measure the power output of her father's car by adapting the method she used in problem M15 with her bicycle. She places a friction belt over the engine flywheel and connects each end to a newton balance as before. With the engine turning the flywheel at 3000 r.p.m. the newton balances register 1200 N and 800 N.
(a) What is the value of the frictional force between the belt and the flywheel?
(b) The circumference of the flywheel is 0.40 m. Find the work done by this force in 1 revolution of the flywheel.
(c) Calculate the useful power output of the car engine.

17. (M16)
(a) (i) What is the frictional force acting on a trolley of mass 2.0 kg as it rolls down a 30° slope at constant speed?
(ii) What is the frictional force acting upon the same trolley if it accelerates down a 30° slope at 3.0 m s^{-2}?
(b) The trolley in part (a) is fitted with an electric motor so that its new mass is 2.2 kg. It then moves up a 30° slope at a steady speed of 0.25 m s^{-1}. The frictional force acting upon it is 13 N. What is the power output of the electric motor?

UNIT H—Heat and kinetic theory of gases

Worked Examples

H1 Calculating from a temperature-time graph
A 50 W electric heater is used by a pupil to heat up a solid of mass 0.500 kg in a well-insulated container. The temperature-time graph for the heating process is shown below. All of the substance has changed into a gas 800 seconds after switching on the heater.

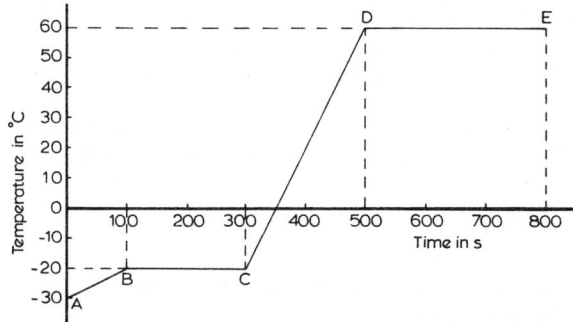

(a) Give a brief description of what is happening in each section **AB**, **BC**, **CD** and **DE**. (2)
(b) Find (i) the specific heat capacity of the solid, (ii) the specific heat capacity of the liquid, (iii) the specific latent heat of fusion, (iv) the specific latent heat of vaporisation. (8)

H2 Specific latent heat
Mr MacLaggan sets up the following apparatus during a physics lesson. He puts cold water into a well-insulated polystyrene container and then injects dry steam to heat up the water. From his results, he hopes to get his pupils to calculate a value for the specific latent heat of vaporisation of water.

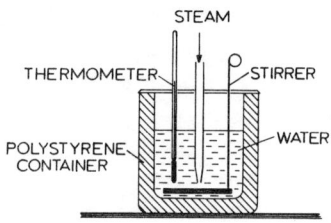

The following information is recorded on the overhead projector:
Temperature of water before steam injection = 24° C
Temperature of water after steam injection = 50° C
Mass of water before steam injection = 0.500 kg
Mass of water after steam injection = 0.521 kg

(a) How much steam does Mr MacLaggan inject into the water? (1)
(b) How much energy is given to the cold water? (2)
(c) How much energy is released as the water formed from steam drops from 100° C to 50° C? (2)
(d) By comparing (b) and (c), deduce how much energy is released when the steam condenses. (2)
(e) What value should his class find from his results for the specific latent heat of vaporisation of water? (3)

H3 The gas laws

(a) A gas is allowed to expand at a constant temperature of 27° C until its volume has doubled. Then it is heated at constant volume until its pressure returns to its original value. These two changes are shown on the following p-V graph as **a** to **b**, and then **b** to **c**.

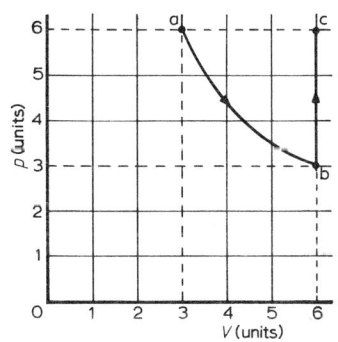

(i) State the gas law which applies over part **ab** of the graph.
(ii) State the gas law which applies over part **bc** of the graph.
(iii) Calculate the temperature of the gas corresponding to position **c**. (6)
(b) Draw the corresponding graph of pressure against temperature for the two changes described in part (a). (2)
(c) Using a kinetic theory model of gases, give a molecular explanation of (i) the reduction in pressure at constant temperature (**ab**) and (ii) the increase in pressure at constant volume (**bc**) which are shown in the graph of part (a). (2)

H4 Introducing the kinetic theory of gases

In this problem, we are looking in on molecular motion in an ideal gas. A molecule of mass m is travelling between points A and B which are centres of opposite faces of a cubical container of side L, making perfectly elastic collisions with the walls at A and B.

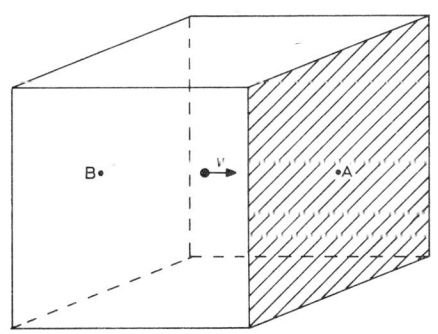

(a) What momentum is imparted to the wall at A when the molecule collides with it? (1)
(b) Show that the number of collisions which this molecule makes with A in unit time is $\dfrac{v}{2L}$. (1)
(c) What is the rate at which momentum is imparted to the wall at A, i.e. the force on the wall due to a single molecule? (1)
(d) If N molecules are moving randomly within the cubical container, show that the pressure on the shaded wall is given by

$$p = \frac{\frac{1}{3} N m \overline{v^2}}{V}$$

where V is the volume of the cube and $\overline{v^2}$ is the mean square speed of the molecules. (4)
(e) The square root of the mean square speed is called the root mean square speed of the molecules and is usually denoted $v_{\text{r.m.s.}}$.
Show that

$$v_{\text{r.m.s.}} = \sqrt{\frac{3p}{\rho}}$$

where ρ is the density of the gas. (3)

H5 Problems on the kinetic theory of gases

According to the gas laws derived from experiment, the pressure, volume and temperature of a gas are linked as follows:

$$\frac{pV}{T_K} = \text{constant, for a fixed mass of gas.}$$

According to the kinetic theory of gases, the pressure and volume are linked to the molecular properties by

$$pV = \tfrac{1}{3} N m \overline{v^2}$$

(a) Show that, for the theory and experiment to be compatible, the average kinetic energy of the molecules and the Kelvin temperature of the gas must be directly proportional to each other. (3)
(b) A particular gas sample has $v_{\text{r.m.s.}} = 200$ m s^{-1} when the temperature is 27° C. At what temperature would the gas molecules have $v_{\text{r.m.s.}} = 400$ m s^{-1}? (3)
(c) Compare the total kinetic energy of the gas molecules in the two states mentioned in (b). (2)
(d) If a gas has a density of 1.25 kg m^{-3} at s.t.p., what is the root mean square speed of its molecules? (2)

H6 Compressing a gas

A gas G is trapped in a tube by a piston P. The tube has a diameter of 0.04 m. When the piston is 0.25 m from the end, the pressure of the gas G is 2.0×10^5 Pa (N m^{-2}).

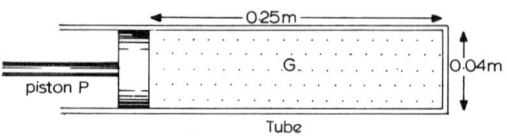

The root mean square speed of the gas molecules ($v_{r.m.s.}$) is 400 m s^{-1} and the gas temperature stays constant.
(a) Calculate the mass of gas G in the tube. (3)
(b) The piston is now pushed slowly into the tube until it is 0.20 m from the end. It is pushed *slowly* to prevent the temperature of the gas rising.
  (i) What is the new pressure of the gas?
  (ii) What is the root mean square speed of the molecules at this stage? Give reasons for your answer. (5)
(c) If the piston was allowed to return *rapidly* to its original position, what would you expect to happen to
  (i) the gas temperature?
  (ii) the value of $v_{r.m.s.}$? (2)

Solutions to Worked Examples

H1
(a) *AB* : Solid is being heated up at a steady rate from $-30°$ C to $-20°$ C.
   BC : Solid is melting. Energy supplied from the heater does not increase the temperature of the substance but does change its state.
   CD : Liquid is being heated up uniformly from $-20°$ to $60°$ C.
   DE : Liquid is changing into gas at the boiling point ($60°$ C). Energy supplied by the heater is used to change the state of the substance from liquid to gas.

(b) (i) s.h.c. of solid $= c = \dfrac{E}{m \Delta T} = \dfrac{50 \times 100}{0.500 \times 10} = 1000$ J kg^{-1} K^{-1}

   (ii) s.h.c. of liquid $= c = \dfrac{E}{m \Delta T} = \dfrac{50 \times 200}{0.500 \times 80} = 250$ J kg^{-1} K^{-1}

   (iii) $l_f = \dfrac{E}{\Delta m} = \dfrac{50 \times 200}{0.500} = 2.0 \times 10^4$ J kg^{-1}

   (iv) $l_v = \dfrac{E}{\Delta m} = \dfrac{50 \times 300}{0.500} = 3.0 \times 10^4$ J kg^{-1}

H2
(a) mass of steam = increase in mass of water = 0.021 kg
(b) $E \uparrow = cm \Delta T \uparrow = 4200 \times 0.500 \times (50 - 24) = 54\,600$ J
(c) $E \downarrow = cm \Delta T \downarrow = 4200 \times 0.021 \times (100 - 50) = 4410$ J
(d) Energy released as steam condenses at $100°$ C $= (54\,600 - 4410) = 50\,190$ J
(e) specific latent heat of vaporisation $= \dfrac{\text{energy released (J)}}{\text{mass condensed (kg)}}$

$\Rightarrow l_v = \dfrac{E}{m} = \dfrac{50\,190 \text{ J}}{0.021 \text{ kg}} = 2.4 \times 10^6$ J kg^{-1}

II3
(a) (i) Since the temperature of the gas is kept constant, Boyle's Law applies to change a to b.

pV = constant

The volume of a given mass of gas varies inversely as the pressure applied to it, provided the temperature stays the same.

(ii) Since the volume of the gas is kept constant, the pressure-temperature gas law applies to change **b** to **c**.

$$\frac{p}{T_K} = \text{constant}$$

The pressure of a given mass of gas is directly proportional to its Kelvin temperature, provided its volume stays constant.

(iii) Use the general gas relationship

$$\frac{P_a V_a}{T_{K_a}} = \frac{P_c V_c}{T_{K_c}} \quad \text{(Remember to use Kelvin temperatures).}$$

$$\frac{6 \times 3}{(273 + 27)} = \frac{6 \times 6}{T_{K_c}} \Rightarrow T_{K_c} = 600 \text{ K (or } 327° \text{ C)}$$

(b) Pressure-temperature graph :

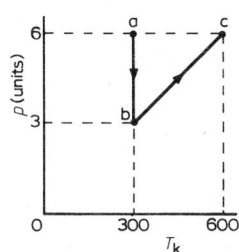

(c) (i) Since the temperature is constant over **ab**, there is no change in the r.m.s. speed of the gas molecules in the sample. However the molecules are occupying an increasing volume and so there are *fewer* collisions with the walls *per second*. Thus the pressure reduces as the volume increases.
(ii) Since the temperature is rising, the r.m.s. speed of the gas molecules is increasing. The molecules arrive at the walls with *more momentum* and pass more momentum to the walls *per collision*. Thus the pressure increases as the temperature increases.

H4

(a) Before the impact, the momentum is $+ mv$ towards point A.
After the impact, the momentum is $- mv$ away from point A.
The momentum of the molecule has changed by $2mv$. Thus the momentum imparted to the wall per impact is $2mv$.
(Remember that momentum is a vector, and that k.e. is conserved).

(b) time taken for molecule to cross between A and B is $\frac{L}{v}$.

time between collisions = time to go from A to B and back to A = $\frac{2L}{v}$.

Therefore the number of collisions per unit time is $\frac{v}{2L}$.

(c) Total momentum imparted by single molecule per unit time is given by

$(2mv) \times \left(\dfrac{v}{2L}\right)$.

$\Rightarrow$ rate of change of momentum $= \dfrac{mv^2}{L} =$ force exerted on wall by single molecule.

(d) total force on wall due to molecules in container $= \left(\dfrac{N}{3}\right) \times \left(\dfrac{\overline{mv^2}}{L}\right)$

The factor $\tfrac{1}{3}$ is needed because approximately $\tfrac{1}{3}$ of the molecules are travelling between the two walls considered. We use $\overline{v^2}$ to take account of the range of speeds in the sample.

now pressure $= \dfrac{\text{force}}{\text{area}} = \dfrac{\left(\dfrac{N}{3}\right) \times \left(\dfrac{\overline{mv^2}}{L}\right)}{L^2} = \dfrac{\tfrac{1}{3} N\overline{mv^2}}{V}$

since volume of box $= V = L^3$.

(e) Since $p = \dfrac{\tfrac{1}{3} N\overline{mv^2}}{V}$, we can write $p = \dfrac{\tfrac{1}{3} M\overline{v^2}}{V} = \tfrac{1}{3} \rho \overline{v^2}$,

where ρ is the gas density $\left(\text{density} = \dfrac{\text{mass}}{\text{volume}}\right)$. Also M is the total mass of the gas sample $(M = Nm)$.

Changing the subject $\Rightarrow \overline{v^2} = \dfrac{3p}{\rho}$ and so $v_{\text{r.m.s.}} = \sqrt{\dfrac{3p}{\rho}}$.

H5

In what follows, $\overline{v^2}$ represents the mean square velocity.

(a) $pV = AT_K$, where A is a constant and T_K is the Kelvin temperature.
$pV = \tfrac{1}{3} N\overline{mv^2} = \tfrac{2}{3} N \times (\tfrac{1}{2} \overline{mv^2}) = \tfrac{2}{3} N \times$ (average k.e. of molecule)
Comparing, we see that $AT_K = B \times$ (average k.e. of molecule), where B is a constant for a fixed sample of gas $(B = \tfrac{2}{3} N)$.

$T_K = \dfrac{B}{A} \times$ (average k.e. of molecule) $=$ constant $\times$ k.e.

(b) Since the average k.e. is proportional to the Kelvin temperature,

$\dfrac{\tfrac{1}{2} m v_1^2}{\tfrac{1}{2} m v_2^2} = \dfrac{T_{K_1}}{T_{K_2}}$ or $\dfrac{T_{K_1}}{T_{K_2}} = \dfrac{v_1^2}{v_2^2} = \dfrac{(200)^2}{(400)^2} = 0.25$

$\Rightarrow T_{K_2} = \dfrac{T_{K_1}}{0.25} = (273 + 27) \times 4 = 1200$ K

(c) The gas sample has four times the kinetic energy at the higher temperature (300 K to 1200 K)

(d) Use the relationship $v_{\text{r.m.s.}} = \sqrt{\dfrac{3p}{\rho}}$ and substitute,

$v_{\text{r.m.s.}} = \sqrt{\dfrac{3 \times 1.0 \times 10^5}{1.25}} = 490$ m s^{-1}

H6

(a) According to the kinetic theory of gases

$$pV = \tfrac{1}{3} Nm (v_{r.m.s.})^2$$

where the symbols have their usual meanings.

$$\text{mass of gas} = Nm = \frac{3pV}{(v_{r.m.s.})^2} = \frac{3 \times 2.0 \times 10^5 \times 0.25 \times \pi \times (0.02)^2}{(400)^2}$$

$\Rightarrow$ mass of gas = 1.18×10^{-3} kg

(b) (i) $p_1 V_1 = p_2 V_2 \Rightarrow p_2 = \dfrac{p_1 V_1}{V_2} = \dfrac{2.0 \times 10^5 \times 0.25}{0.20}$

new pressure $p_2 = 2.5 \times 10^5$ Pa.
(ii) The root mean square speed stays at 400 m s^{-1} because there has been no change in temperature. Thus there has been no change in the average k.e. of the gas molecules.
(c) (i) temperature reduces (ii) $v_{r.m.s.}$ reduces too.

Practice Questions

1. (H1)
Here is some information about the thermal properties of aluminium :
Specific heat capacity = 900 J kg^{-1} K^{-1}
Melting point = 660° C
Specific latent heat of fusion = 4.02×10^5 J kg^{-1}
(a) How much energy would be needed to convert a 5.0 kg block of aluminium, originally at 20° C, into a molten liquid at 660° C?
(b) A 60 W immersion heater is fitted into a solid block of aluminium of mass 3.0 kg and switched on for 5 minutes. Estimate the maximum rise in temperature produced.
(c) When a block of aluminium at 100° C is dropped into a trough of 10 kg of water at 20° C, the temperature of the aluminium drops to 30° C. Estimate the mass of the block. State what assumptions were necessary to obtain your answer.
(Specific heat capacity of water = 4200 J kg^{-1} K^{-1})

2. (H1)
Here is a chart giving the thermal properties of three substances :

Substance	Melting point (°C)	Boiling point (°C)	s.h.c. of solid (J kg^{-1} K^{-1})	s.l.h. of fusion (J kg^{-1})	s.h.c. of liquid (J kg^{-1} K^{-1})	s.l.h. of vaporisation (J kg^{-1})
A	100	200	2.0×10^2	8.0×10^5	8.0×10^2	2.0×10^5
B	200	300	4.0×10^2	4.0×10^5	4.0×10^2	4.0×10^5
C	300	400	8.0×10^2	2.0×10^5	2.0×10^2	8.0×10^5

Each substance has a mass of 1.0 kg and starts at 0° C. Energy is given to each substance at the same rate by identical heaters.
(a) Which substance remains longest in the solid state?
(b) Which substance remains longest in the liquid state?

(c) Which substance requires the most energy at the melting point to turn from a solid into a liquid?
(d) Which substance requires the most energy at the boiling point to turn from a liquid into a gas?
(e) Calculate the total amount of energy which has to be supplied to substance B to convert it completely into a gas.

3. (H2)
0.500 kg of water is poured into an insulated beaker which is placed on the top pan of a balance. The water is then heated by a 400 W immersion heater and the reading on the balance is recorded as shown :

Time (in minutes)	0-7	7-17
Balance reading (kg)	stays at 0.70	drops steadily from 0.70 to 0.60

(a) Calculate the starting temperature of the water, assuming that the s.h.c. of water is 4.2×10^3 J kg^{-1} K^{-1}.
(b) Calculate a value for the specific latent heat of vaporisation of water using the above information.

4. (H3)
On a day when the atmosphere pressure is 1.0×10^5 Pa and the temperature is 7° C, the pressure in a car tyre is raised to 1.50×10^5 Pa *above the pressure of the atmosphere*.
(a) Assuming that there are no leaks in the tyre, what will the absolute gas pressure in the tyre be when its temperature is raised to 27° C during a long journey?
(b) While at 27° C, the tyre develops a small puncture and the pressure in it drops by 5.0×10^4 Pa. What percentage of the molecules have escaped due to the puncture?
(c) What assumptions have you had to make about the tyre to obtain answers to parts (a) and (b)?

5. (H4)
A beam of molecules, each of mass 6.4×10^{-27} kg, travels horizontally at 5.0×10^2 m s^{-1} towards a vertical wall of a box. The molecules bounce straight back off the wall and then move in the opposite direction with exactly the same speed as before.
(a) Explain why this type of collision is referred to as a 'perfectly elastic' collision.
(b) One million such particles collide with the wall each second. What force do they exert on the wall?
(c) The cross sectional area of the beam is 2.0×10^{-6} m^2. What pressure do the particles exert on this part of the wall?

6. (H6)
(a) A spherical gas tank of internal diameter 6.0 m contains 400 kg of gas. The root mean square speed of the gas molecules is 600 m s^{-1}. Calculate the pressure of the gas in the tank.
(b) The pressure of the gas in the tank is adjusted to 8.0×10^5 Pa, the volume and the mass of the gas being the same as before. Calculate the new root mean square speed of the gas molecules.

UNIT E—Electricity

Worked Examples

E1 Introducing the capacitor
A and B represent two uncharged brass discs with insulated handles. The discs are parallel to each other and a distance d apart.

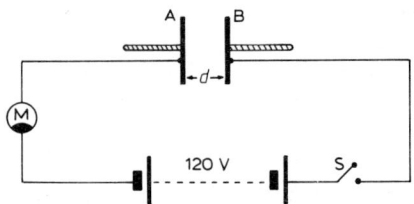

(a) When switch S is closed, the needle on meter M moves quickly across the scale and then returns to zero, indicating that charge has been transferred from the battery to the discs.
   (i) Explain why disc A has become negatively charged whereas disc B has become positively charged. (2)
   (ii) Why does the charge transfer stop after a very short period of time? (2)
   (iii) What is the p.d. across the discs after charging? (1)
(b) The readings on M are recorded with the discs placed at various distances d apart, starting with both discs uncharged each time. Here is a table of the results obtained :

Disc separation, d (mm)	8.0	6.0	4.0	2.0
Charge transferred, q (units)	3.0	4.0	6.0	12.0

   (i) Explain in terms of induction why the discs store more charge when they are closer together. (2)
   (ii) What name is given to a device which is used to store charge? (1)
   (iii) Apart from bringing the discs closer together, what else could be done to increase the charge stored on each plate? (2)

E2 Factors affecting capacitance
(a) A pupil uses the apparatus described in the previous question to find out how the amount of charge stored on each plate (disc) of this basic capacitor depends upon various factors. He summarises his results as follows :

Separation of discs, d ↑	Charge stored, Q ↓	
Overlap area of discs, A ↑	Charge stored, Q ↑	$Q = \dfrac{kAV}{d}$
p.d. across discs, V ↑	Charge stored, Q ↑	
Insert dielectric, k	Charge stored, Q ↑	

(i) What exactly does the pupil mean by 'overlap area of discs'? (2)
(ii) Explain what is meant by the term 'dielectric' and give two examples of suitable materials. (2)
(iii) Why must the discs not be brought *too* close to each other? (2)
(b) The two-plate device discussed in (a) is used for storing charge and is called a **capacitor**. Its ability to store charge is called its **capacitance**, and is defined as follows :

$$\text{capacitance} = \frac{\text{charge stored on each plate of the capacitor}}{\text{p.d. across the plates of the capacitor}}$$

$$C \text{ (in farads)} = \frac{Q \text{ (in coulombs)}}{V \text{ (in volts)}}$$

(i) Which of the following equations is correct for a parallel-plate capacitor?

$$C = \frac{kA}{d}, \quad C = \frac{kd}{A}, \quad C = kAd. \quad (2)$$

(ii) If a capacitor stores 2.4×10^{-3} coulombs of charge on each plate when a 100 V battery is connected across it, what is its capacitance in microfarads? (2)

E3 The *R-C* charging circuit
When switch S is closed, the capacitor C charges up slowly through a high value resistor R. The charging current i is recorded on the microammeter.

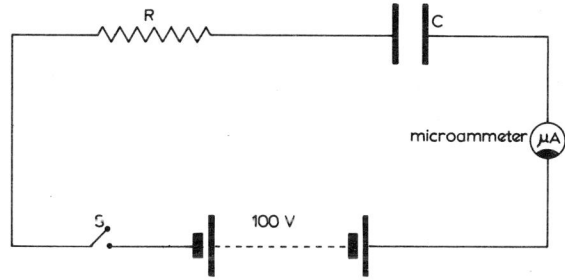

Here is the graph of charging current against time, taken from the instant of switching :

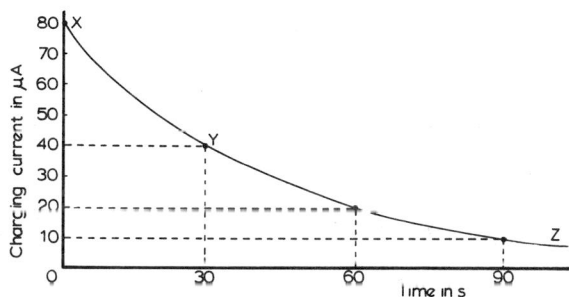

(a) Explain why the charging current drops towards zero as time goes on. (4)
(b) What time interval is required for the current to reduce from 4 µA to 1 µA? (2)

(c) What would happen to the shape of the graph for each of the following changes?
  (i) A resistor of twice the resistance is used instead. (2)
  (ii) A capacitor of twice the capacitance is used instead. (2)

E4 Charging and discharging a capacitor
(a) In the following circuit, a single pole, two-way switch S is used so that the capacitor can be charged (S-1) or discharged (S-2).
The p.d.s across R and C are recorded for each setting of the switch and graphs of voltage against time are drawn. One graph is given. Complete the other three graphs for the switch connections indicated. (6)

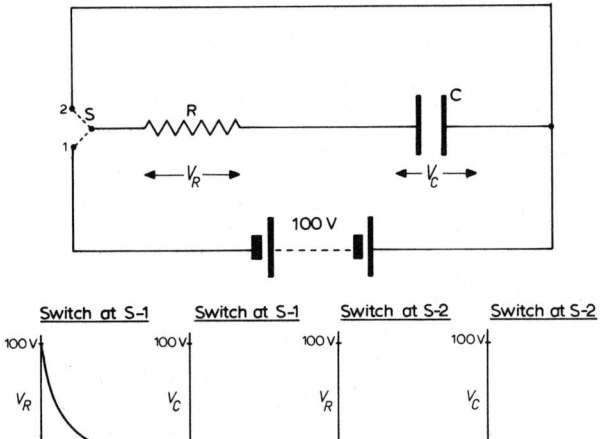

(b) Given the following values : R = 4.7 MΩ and C = 8.0 µF, calculate
  (i) the maximum value of the charging current when S is at S-1,
  (ii) the maximum charge stored by C when S is at S-1. (4)

E5 Millikan's oil drop experiment
In a Millikan's Oil Drop experiment, two conducting plates are placed horizontally 2.0 cm apart and a p.d. of 400 V is applied across them. The diagram shows the electric field pattern between the plates.

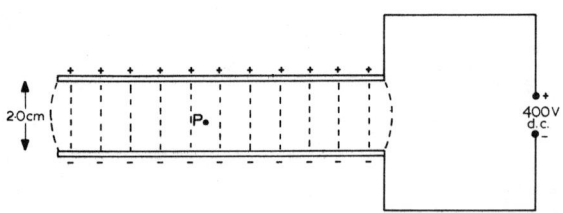

(a) Discuss briefly how it is possible to hold a charged oil drop *stationary* at position P between the plates. (2)
(b) The upwards force F_E acting on the charged oil drop due to the electric field is given by $F_E = \dfrac{qV}{d}$, where q is the charge on the oil drop, V is the p.d. across the

plates and d is the separation of the plates. Use this to calculate the charge carried by a stationary oil drop of mass 6.4×10^{-16} kg. (2)

(c) In Millikan's experiment, the charge carried is calculated for a large number of oil drops. Typical results are shown below :

Oil drop	A	B	C	D	E	F	G	H
Charge carried ($\times 10^{-19}$ C)	+3.2	−6.4	−1.6	+6.4	+9.6	+1.6	−3.2	+6.4

What conclusion regarding the nature of charge can be drawn from this set of results? (3)

(d) During another stage of the experiment, oil drops were observed travelling between the plates at *constant velocities*. Discuss the forces acting upon the oil drop when such a motion is obtained. (3)

E6 Combining resistors

(a) What is the value of resistor X if the effective resistance between terminals A and B is $10 \, \Omega$? (3)

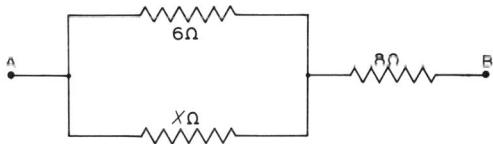

(b) What is the value of resistor Y if the effective resistance between terminals C and D is $4 \, \Omega$? (3)

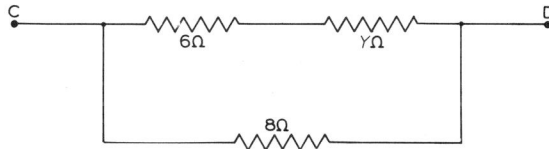

(c) 60 V d.c. supplies are connected across AB and CD. Calculate the currents flowing through resistors X and Y. (4)

E7 Shunts and multipliers

In the diagram, M represents a moving coil galvanometer whose coil has a resistance of $2 \, \Omega$. This basic meter gives a full scale deflection when a current of 2 mA is passed through its coil.

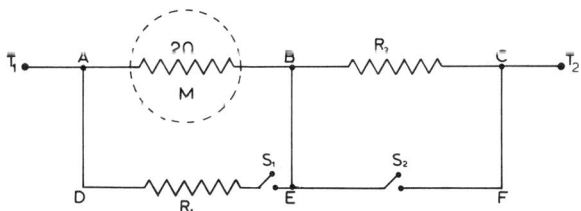

48 Electricity

When both switches are open, the instrument operates as a **voltmeter** with a range of 0 - 1 V, and when both switches are closed, it operates as an **ammeter** with a range of 0 - 1 A.
(a) Describe how the instrument can operate (i) as a voltmeter, and (ii) as an ammeter. (4)
(b) Calculate the values of the resistors R_1 and R_2. (4)
(c) What is the current in R_1 when the ammeter registers 0.5 A? (1)
(d) What is the current in R_2 when the voltmeter registers 0.5 V? (1)

E8 Introducing internal resistance

In the circuit below, R_{INT} represents the internal resistance of a cell and R_{EXT} represents the external resistance of the circuit. A high resistance voltmeter is placed across the terminals AB of the cell and a low resistance ammeter is used to record the circuit current.

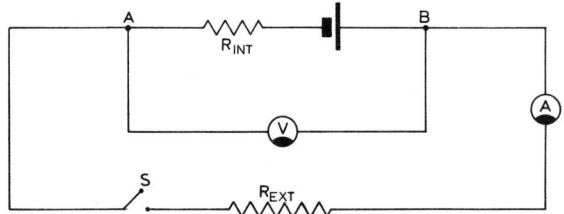

When S is open, V registers 2.0 V. When S is closed, it registers 1.6 V and the ammeter registers 0.8 A.
(a) What is the e.m.f. of the cell, in volts? (1)
(b) What is the terminal potential difference when S is closed? (1)
(c) Calculate the values of R_{INT} and R_{EXT}. (4)
(d) What would be the new readings on V and A if the external resistance were halved? (2)
(e) Explain what is meant by the term 'lost volts'. (2)

E9 Problems on internal resistance

Bill needs a 1.5 V cell to operate his torch. He checks the p.d. across the terminals of a cell and obtains a reading of 1.5 V on the voltmeter (Fig. I). However, when he puts the torch bulb across the cell, the voltmeter reading drops to 1.2 V and the bulb is not very bright (Fig. II).

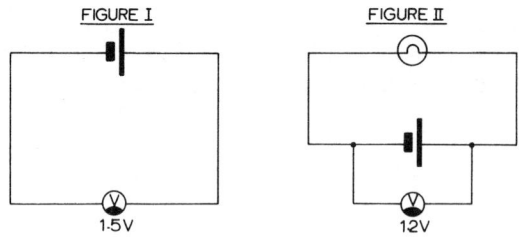

(a) Explain why the reading on the voltmeter drops whenever the bulb is inserted. (3)
(b) His friend says that he should use a 1.5 V cell which has a lower internal resistance. Why would this improve things? (3)

(c) The current flowing in Fig. II is 0.3 A. Calculate
   (i) the resistance of the torch bulb (R_{EXT}). (2)
   (ii) the internal resistance of the cell (R_{INT}). (2)

E10 The Wheatstone bridge
Tom, Dick and Barry are asked to construct balanced Wheatstone Bridges. The circuits which they produce are shown below.

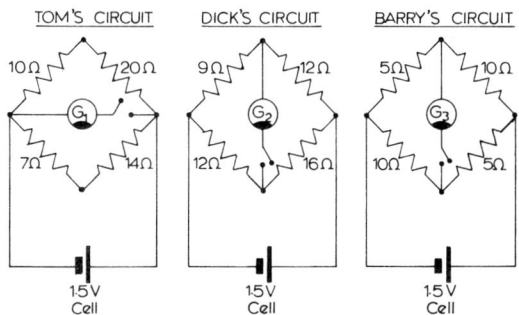

One of the circuits gives a balanced Wheatstone Bridge, one gives an off-balance Wheatstone Bridge and one is definitely not recommended!
(a) Whose circuit is definitely not recommended? Explain why. (3)
(b) Whose circuit gives a balanced Wheatstone Bridge? Explain why. How would you test that balance had been obtained? (3)
(c) In the unbalanced Wheatstone Bridge, explain which way the electron current will flow through the galvanometer when the switch is closed. (4)

E11 The metre bridge
In the following metre bridge circuit, AB is a uniform resistance wire of length 100 cm, G is a sensitive, centre-zero galvanometer and R_S is a standard resistor of value 6.00 Ω. When a resistor R_U is in position, balance is obtained when the slider is 40 cm from end A of the wire.

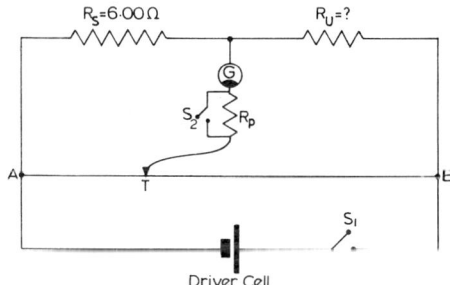

(a) Explain why the galvanometer must be sensitive and have a centre zero. (2)
(b) What are switches S_1 and S_2 for? (2)
(c) Find the value of the unknown resistor R_U. (2)

50 Electricity

(d) What would happen to the balance position AT in each of the following modifications?
 (i) The terminals of the driver cell are fitted the opposite way across wire AB.
 (ii) Another identical standard resistor is connected in parallel with the original one. (4)

E12 Introducing the potentiometer

In this potentiometer circuit, the uniform resistance wire AB is 1.00 m long and G is a sensitive, centre-zero galvanometer. The two-way selector switch S introduces a standard cell of e.m.f. 1.10 V (E_S) when in one position, and replaces it with a cell of unknown e.m.f. (E_U) when in the other position.

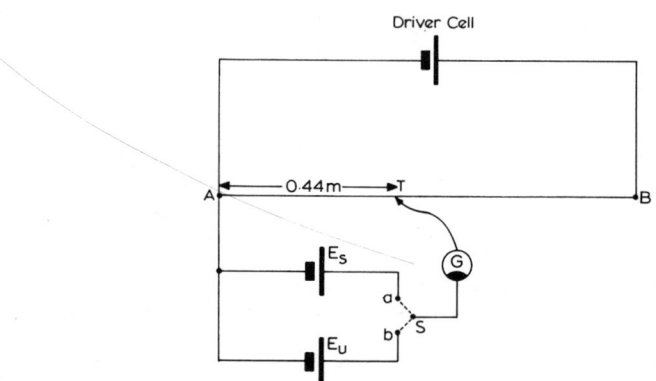

With S in position a, the reading on G becomes zero when AT is set to 0.44 m.
With S in position b, the reading on G becomes zero when AT is set to 0.80 m.
(a) Calculate the potential gradient along the wire AB. (2)
(b) What is the e.m.f. (E_U) of the unknown cell? (2)
(c) Wire AB is nichrome wire of uniform cross-section and resistance 25 ohms per metre. How much current is delivered by the driver cell at balance? (2)
(d) Explain why it is customary to choose a potentiometer wire which has
 (i) a uniform cross-section and (ii) a substantial resistance per metre. (2)
(e) Very often, the following modification is made to the detecting part (TS) of the circuit :

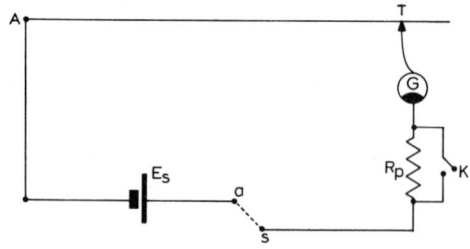

Explain the function of R_p and K and describe how they improve the operation of the potentiometer. (2)

E13 The potentiometer – problems

(a) A pupil designs the following circuit for a potentiometer experiment.

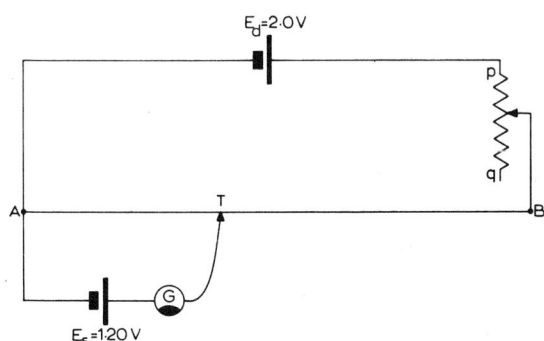

She records the relevant information in her laboratory notebook:

Length of potentiometer wire AB	5.00 m
Resistance of wire used	12 Ω m^{-1}
Range of rheostat pq	0 - 20 Ω
e.m.f. of driver cell, E_d	2.0 V
e.m.f. of standard cell, E_s	1.20 V

Use this information to prove that the balance length AT increases from 3.00 m to 4.00 m when the rheostat slider is moved from **p** to **q**. (6)

(b) The pupil then rearranges the circuit as shown and obtains a new balance at AT = 2.50 m.

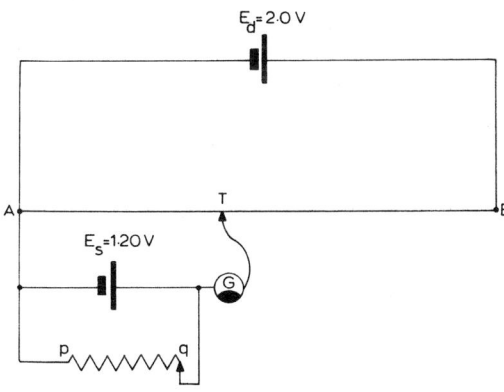

What is the internal resistance of the standard cell? (4)

E14 Electromagnetism

(a) This is a simplified diagram of a loudspeaker. The speech coil is attached to the neck of the loudspeaker cone. A d.c. supply is connected across terminals A and B of the electromagnet, whereas an a.c. supply is connected across terminals C and D of the speech coil.

52 *Electricity*

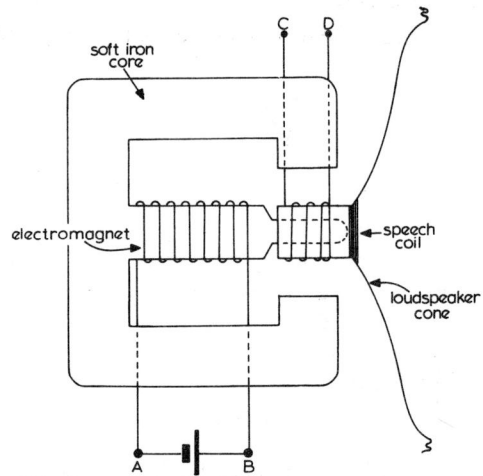

(i) Explain why the cone vibrates at the frequency of the a.c. supply. (3)
(ii) Which way, in or out, will the cone move when electrons are moving through the speech coil from C to D? Justify your choice. (3)
(b) A wire is stretched between the opposite poles of a permanent magnet and an alternating current is passed through it at a frequency of 50 Hz.

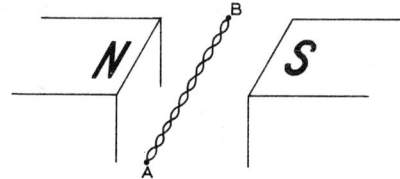

(i) Explain why a set of standing waves is produced in the wire. (2)
(ii) Estimate the speed at which waves travel through this wire, given that AB = 60 cm. (2)

E15 Self-inductance
(a) The following circuit contains two identical bulbs, a coil and a resistor.

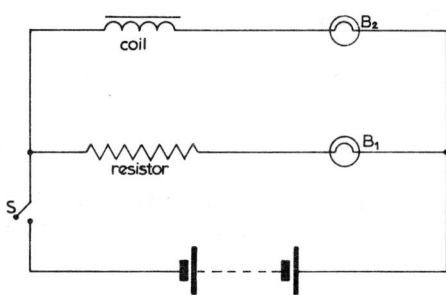

(i) When switch S is closed, bulb B_1 lights up slightly before bulb B_2. Both bulbs reach the same level of brightness. Explain these effects. (4)
(ii) The graph below shows how the currents through the bulbs vary with time.

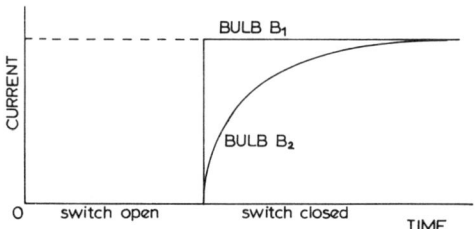

The value of R is increased and the iron core is removed from the coil. Draw the current-time graph to represent this new situation. (3)

(b) Major Disaster has bought himself a new train set which has a step-down transformer to reduce the mains voltage from 240 V a.c. to 12 V a.c. The Major finds a 240 V d.c. supply amongst his wartime souvenirs and he uses it across the primary of the transformer. The train refuses to move and after a short while smoke starts pouring out of the transformer. His superior officer, General Alert, is able to explain the lack of motion and the clouds of smoke. Can you? (3)

E16 Revising the transformer

The bulbs in the following circuit are each rated '12 V, 24 W'. When switched on, they operate at their correct rating. The transformer is 100% efficient.

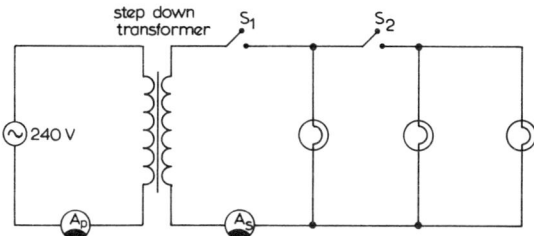

(a) Calculate the turns ratio of the transformer. (1)
(b) What readings would be obtained on the a.c. ammeters A_p and A_s when (i) S_1 only is closed? (ii) S_1 and S_2 are both closed? (3)
(c) A visiting teacher from the United States brings a cineprojector rated '115 V, 5 A' over to a part of Britain where the mains supply voltage is 230 V. Two pupils suggest the following ways of getting his cineprojector to operate correctly :

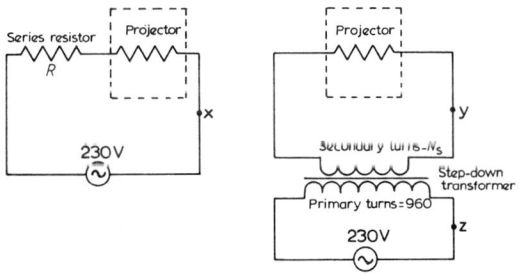

54 Electricity

(i) Calculate the required value of R, in ohms. (2)
(ii) Calculate the value of N_s, the number of secondary turns. (1)
(iii) Compare the efficiencies of the two circuits. (2)
(iv) At which point, x, y or z, is the current smallest? (1)

E17 Electrical power transmission

The diagram represents a simplified version of a modern national grid system. The transformers are assumed to be 100% efficient, for ease of calculation.

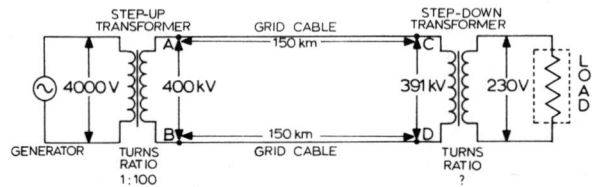

When the generator is producing 600 kW at 4 kV, the load is receiving power at 230 V.
(a) Calculate the current in (i) the generator circuit, (ii) the grid cables and (iii) the load circuit. (4)
(b) What is the resistance per km of the grid cable? (2)
(c) What is the turns ratio of the step-down transformer? (1)
(d) Transformers like the ones used here are very expensive items. A pupil studying economics, but not physics, asks :
"Would it not save a lot of money to step down the generator voltage from 4000 V to 230 V straight away and then transmit the power along the cables directly to the load?" How would you answer this question? Produce figures to back up your answer. (3)

E18 Electrons in thermionic diodes

A demonstration diode has two flat metal electrodes and a heater filament enclosed in a vacuum tube.

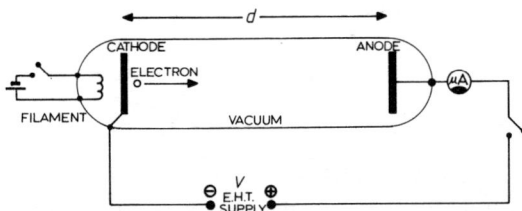

Both switches are closed and a small current is recorded on the microammeter.
(a) When an electron is emitted from the heated cathode, it is accelerated across to the positively charged anode by a horizontal force given by the equation :

$$F = \frac{qV}{d}$$

(i) Show that the kinetic energy of the electron is increased by an amount qV as it travels from cathode to anode. (2)

(ii) Show that the electron increases its speed by an amount $\sqrt{\dfrac{2qV}{m}}$, where m is

the mass of the electron. (2)
(iii) Estimate the speed increase of the electron, given the following information :
Setting of e.h.t. supply, $V = 4.5 \times 10^3$ V
Electronic charge, $q = 1.6 \times 10^{-19}$ C
Mass of electron, $m = 9.0 \times 10^{-31}$ kg (2)
(b) If the reading on the microammeter is 2.4×10^{-6} A, estimate the number of electrons emitted each second from the heated cathode. (4)

E19 The diode and rectification

In this circuit, an ideal diode is connected in series with two resistors of values 2 Ω and 3 Ω. Switch S is moved at a steady rate backwards and forwards between contacts x and y. The e.m.f. of each cell is 1.0 V.

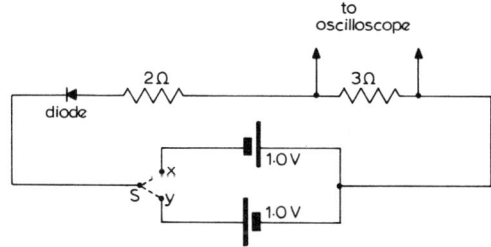

When the p.d. across the 3 Ω resistor is applied to the Y-plates of an oscilloscope, the following trace is produced on the screen :

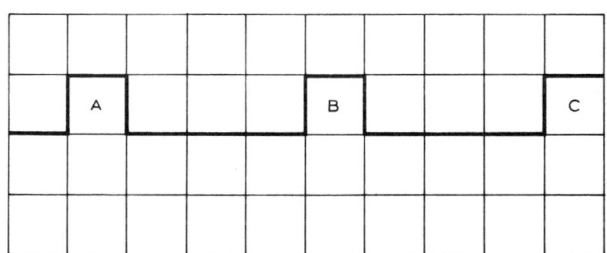

EACH SQUARE IS 1cm x 1cm

The time base setting of the oscilloscope is 10 ms cm^{-1} and the Y-gain setting is 0.6 V cm^{-1}.
(a) Explain why the trace looks like this. (3)
(b) Calculate the frequency of the vibrating switch. (3)
(c) What would be the effect upon the trace of reversing the diode? (1)
(d) What would be the effect upon the trace of connecting the oscilloscope across the 2 Ω resistor, instead of the 3 Ω resistor? (1)
(e) Draw the trace which is obtained when an a.c. supply is fitted instead of the cell-switch arrangement. (2)

E20 Measurements from the oscilloscope

The time-base control of an oscilloscope is set at 10 milliseconds per cm (10 ms cm^{-1}). During experiment A, on the speed of sound, two blips are obtained on the screen, as in TRACE A. During experiment B, on the frequency of a tuning fork, ten waves are obtained on the screen, as in TRACE B.

56 Electricity

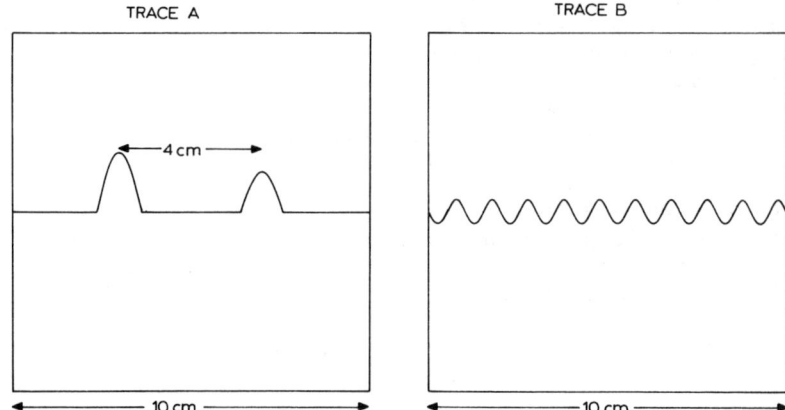

(a) Draw a labelled diagram of the equipment you would use in connection with experiment A. Explain briefly how you would find the speed of sound using this apparatus. (4)
(b) Calculate the frequency of the tuning fork used in experiment B. (3)
(c) What happens to the wave trace when the tuning fork is moved nearer to the microphone? (1)
(d) The oscilloscope control is re-set at 1 ms cm^{-1}. What effect would this have on (i) TRACE A? (ii) TRACE B? (2)

E21 Introducing a.c. circuits

Figures 1 and 2 represent circuits which can be used to compare the brightness of a bulb when it is operated from d.c. and a.c. supplies. The potential dividers are adjusted until the bulb has the *same brightness* for each supply and the p.d. across the bulb is then examined on an oscilloscope.

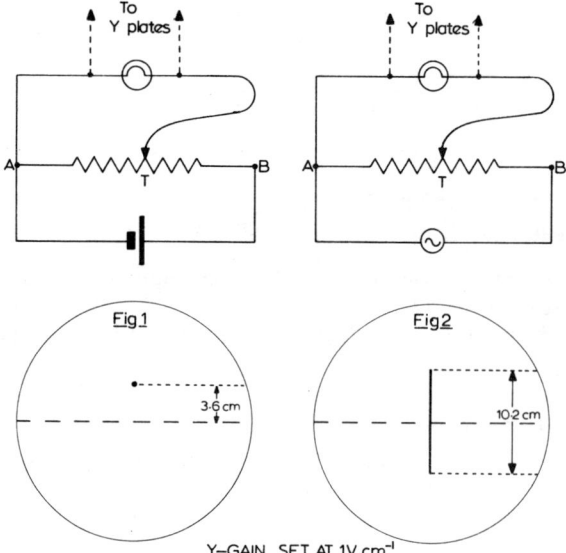

(a) Explain briefly how the potential divider allows the bulb brightness to be adjusted. (2)
(b) What additional equipment would be needed to *test* that the bulb was indeed equally bright on each supply? (2)
(c) How does this experiment help to develop the idea of 'root mean square' voltage? (2)
(d) According to the information obtained from the oscilloscope traces, how is the root mean square voltage related numerically to the peak voltage? (2)
(e) Draw diagrams to indicate what will happen to the traces when the oscilloscope time base is switched on. (2)

E22 The coil in an a.c. circuit

This circuit shows a signal generator connected across a coil and a bulb. When the output of the signal generator is set at 6.0 V r.m.s., 1000 Hz, the bulb is operating at its normal brightness.

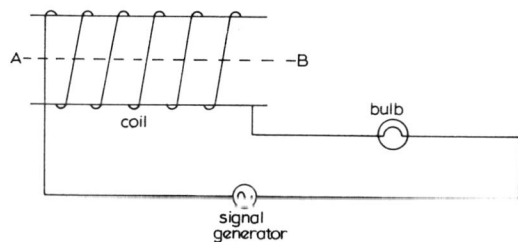

(a) Explain why the bulb goes dimmer when a soft-iron core is fitted along the axis of the coil (AB). (3)
(b) Explain how the bulb brightness can be changed by altering the frequency setting of the signal generator. (2)
(c) Would the bulb be dimmer if the output of the generator were set at 6.0 V r.m.s., 900 Hz or at 6.0 V r.m.s., 1100 Hz? (1)
(d) What name is given to a coil which is used to control the size of the alternating current in circuits? (1)
(e) What name is given to the opposition to a.c. presented by a coil? (1)
(f) Explain why the coil is wound using thick copper wire. (1)
(g) What is the peak voltage used in this experiment? (1)

E23 Frequency response of an inductor

An inductor L and an a.c. milliammeter are connected across a 12.0 V r.m.s. variable frequency supply.

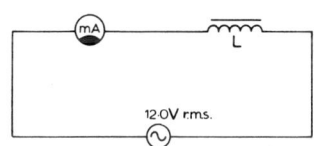

The circuit current is recorded at various frequencies. The results are shown below.

Frequency in Hz	100	200	300	400	500
Current in mA	?	6.0	4.0	3.0	?

The inductive reactance of the inductor is a suitable measurement of the opposition to a.c. presented by the inductor. It is defined as follows:

inductive reactance (Ω) = $\dfrac{\text{r.m.s. voltage across the inductor (V)}}{\text{r.m.s. current through the inductor (A)}}$

$$X_L = \dfrac{V_L}{I_L}$$

(a) What would you expect the inductive reactance to be at (i) 100 Hz and (ii) 500 Hz? (4)
(b) What is the peak current flowing in the circuit at 300 Hz? (1)
(c) Draw graphs of (i) current I_L against frequency f, and (ii) inductive reactance X_L against frequency f. (3)
(d) The soft-iron core is now removed from the inductor. What effect would this have on (i) the value of the current? (ii) the value of the inductive reactance? (2)

E24 The capacitor in an a.c. circuit

This circuit shows a capacitor and a bulb connected in series across a signal generator. When the output of the signal generator is set at 6.0 V r.m.s., 1000 Hz, the bulb lights up to its normal brightness.

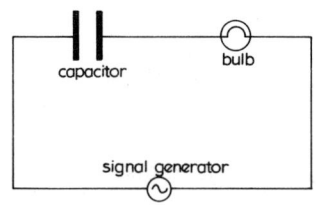

(a) How is it possible for the bulb to light up when the plates of the capacitor are known to be separated by a **non-conducting** material called a dielectric? (3)
(b) Explain why the brightness of the bulb increases when the frequency setting of the signal generator is increased. (2)
(c) What would happen to the bulb brightness if a larger value of capacitance were used? Explain your answer. (2)
(d) Draw an accurate voltage–time graph for the output of the signal generator corresponding to the setting 6.0 V r.m.s., 1000 Hz. (3)

E25 Frequency response of a capacitor

A capacitor and an a.c. milliammeter are connected in series across a 12.0 V r.m.s., variable frequency signal generator. The milliammeter records the r.m.s. value of the current in the circuit.

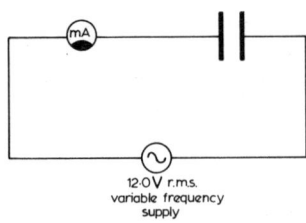

The circuit current is recorded at various frequencies as follows :

Frequency in Hz	100	200	300	400	500
Current in mA	?	32	48	64	?

The capacitive reactance of the capacitor is a suitable measurement of the opposition to alternating current presented by the capacitor. It is defined as follows :

$$\text{capacitive reactance } (\Omega) = \frac{\text{r.m.s. voltage across the capacitor (V)}}{\text{r.m.s. current 'through' the capacitor (A)}}$$

$$X_C = \frac{V_C}{I_C}$$

(a) What would you expect the capacitive reactance to be at (i) 100 Hz? (ii) 500 Hz? (2)
(b) Draw graphs of (i) current I_C against frequency f, (ii) capacitive reactance X_C against frequency f. (3)
(c) What effect would each of the following changes, taken separately, have on the value of the circuit current?
   (i) The dielectric is removed from the capacitor plates.
   (ii) The plates of the capacitor are moved farther apart.
   (iii) The overlap area of the plates is decreased.
   (iv) A larger value of capacitance is used instead.
   (v) A larger value of supply voltage is used instead. (5)

E26 Series and parallel resonant circuits

At the end of a successful series of class experiments upon the resonance effects of a.c. circuits, a teacher writes up the following summary on the blackboard :

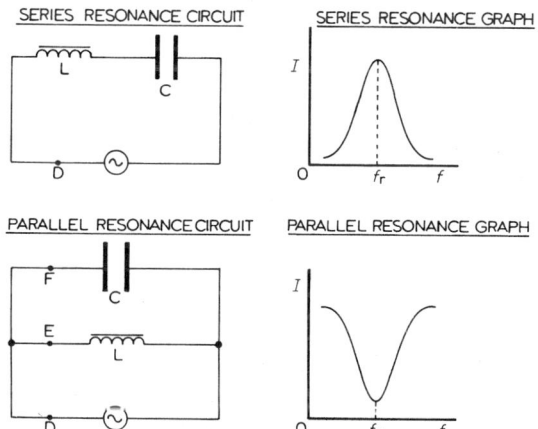

(a) What happens, in each case, to the resonance frequency f_r when the circuits are adjusted in each of the following ways?
   (i) The value of the inductance L is increased by wrapping more turns round the soft-iron core.

(ii) The value of the capacitance C is increased by bringing the plates closer together.
(iii) A resistor is inserted at point D in each circuit. (6)
(b) When the parallel circuit is at resonance, a.c. milliammeters are placed in the inductor and capacitor branches at E and F. The readings are :
current at E = 42 mA, current at F = 42 mA.
At the same time, the current delivered by the supply is almost zero.
How is this possible? (2)
(c) Give one practical application of resonant circuits. (2)

E27 Graphical summary of the a.c. course
During a school course on a.c. circuits, a pupil builds up a set of graphs of current against frequency as shown below.

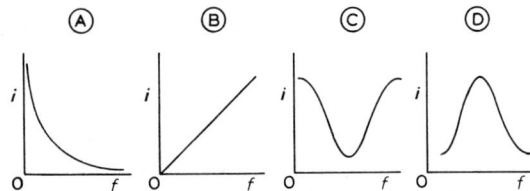

Unfortunately, he forgets to label them as he goes along and is in difficulty when his teacher asks him to draw the four circuits which correspond to the above graphs.
(a) Can you draw the four circuits? (Label them Circuit A, Circuit B, etc.) (6)
(b) Reactance is the term used for the opposition to a.c. Complete the reactance-frequency graphs corresponding to the four graphs above. (4)

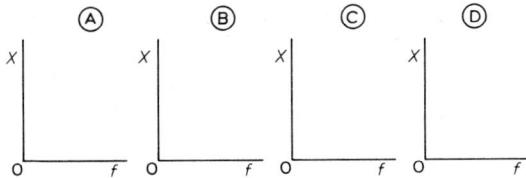

Solutions to Worked Examples

E1
(a) (i) Disc A is connected to the −ve terminal of the battery.
Electrons move from this terminal on to disc A, making it −ve.
Disc B is connected to the +ve terminal of the battery.
Electrons are attracted off disc B by this terminal. Since disc B has lost some electrons, it becomes +ve.
   (ii) Electrons only move on to disc A, and off disc B, for a very short time until the p.d. across the discs equals the p.d. across the battery.
   (iii) 120 V, the p.d. of the battery.
(b) (i) The electrons arriving on disc A set up an electric field which repels electrons from disc B. This field effect will be stronger when the discs are closer together.
   (ii) A **capacitor** is a device for storing charge.
   (iii) A thin sheet of an insulating material (a dielectric) could be fitted in between the discs.

E2
(a) (i) The overlap area is area common to the two plates (shown shaded).

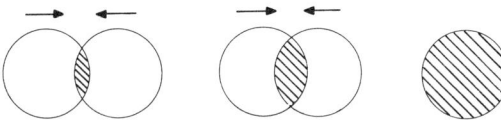

(ii) A sheet of good insulating material fitted between the plates of the capacitor to improve its capacitance. Examples are mica, paraffin wax, ceramics, etc.
(iii) If plates too close, an electric discharge may puncture the dielectric and damage the capacitor.

(b) (i) $C = \dfrac{kA}{d}$ is correct. It agrees with the results shown.

(ii) $C = \dfrac{Q}{V} = \dfrac{2.4 \times 10^{-3} \text{ C}}{100 \text{ V}} = 2.4 \times 10^{-5} = 24 \times 10^{-6}$ F (24 µF)

E3
(a) *Point X on graph*: At the instant of switching, there is no charge on the capacitor and so there is zero p.d. across it. This means that the entire 100 V of the battery must be across the resistor and so a relatively large charging current will flow through the resistor on to the plates of the capacitor.
Point Y on graph: As the capacitor plates become charged a p.d. develops across them, leaving a smaller p.d. across the resistor. Thus a *smaller* charging current will flow through the resistor on to the plates of the capacitor.
Point Z on graph: Eventually the p.d. across the capacitor will get very near to 100 V, leaving virtually zero p.d. across the resistor. Thus the charging current reduces to zero.

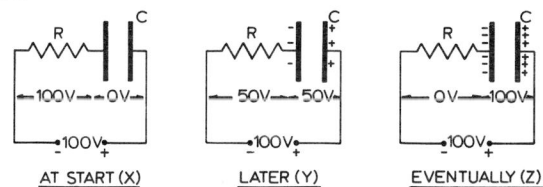

| AT START (X) | LATER (Y) | EVENTUALLY (Z) |

(b) The graph is of a special type called an exponential curve. It has a constant 'half-life' of 30 s, i.e. it takes 30 s for the current to drop from 80µA to 40µA, 40µA to 20µA, 20µA to 10µA, etc. Therefore it will take 60 s (two 'half-lives') to drop from 4µA to 1µA.
(c) (i) Initial value of current would be 40µA instead of 80µA,
'Half-life' would be 60 s instead of 30 s.
(ii) Initial value of current would be 80µA, as before.
'Half-life' would be 60 s instead of 30 s.
To obtain slow charge or discharge rates, use large values of R and C.

E4
(a)

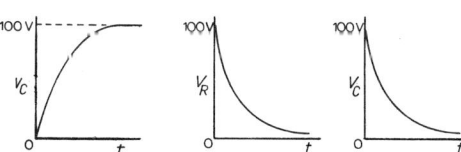

(b) (i) Max. current is when $t = 0$ and is given by Ohm's Law,

$$I_{max.} = \frac{V_R}{R} = \frac{100 \text{ V}}{4.7 \times 10^6 \text{ }\Omega} = 21.3 \times 10^{-6} \text{ A } (21.3 \mu A)$$

(ii) Max. charge is stored when the p.d. across C reaches 100 V.

$Q_{max.} = CV_C = 8.0 \times 10^{-6} \times 100 = 8.0 \times 10^{-4}$ coulombs.

E5

(a) The oil drop can be stationary if the upward force of attraction to the top plate is equal in size to the weight of the oil drop.

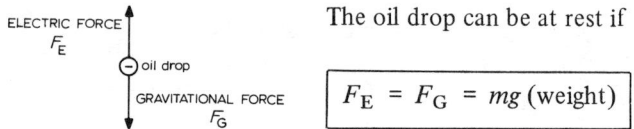

The oil drop can be at rest if

$$\boxed{F_E = F_G = mg \text{ (weight)}}$$

In this case, the top plate is shown +ve, and so the drop would have to be $-$ve to obtain an upwards attraction.

(b) Since $F_E = \dfrac{qV}{d}$, then $q = \dfrac{F_E d}{V} = \dfrac{mgd}{V}$.

Substituting,

$$q = \frac{6.4 \times 10^{-16} \times 10 \times 2.0 \times 10^{-2}}{400} = 3.2 \times 10^{-19} \text{ C}$$

(c) The smallest unit of charge (now known to be the electronic charge) is 1.6×10^{-19} C The charges on the oil drops are all $n \times (1.6 \times 10^{-19} \text{ C})$, where n is a whole number, and is $-$ve if there is a surplus of electrons on the drop, and +ve if there is a shortage of electrons on the drop.

(d) When an oil drop is moving, we have to take into account the effect of air resistance (viscosity). The forces on an oil drop which is moving upwards are :
F_E, the electric force of attraction acting *upwards*,
F_G, the gravitational force acting *downwards*,
F_V, the viscous force acting *downwards*, opposing the motion.

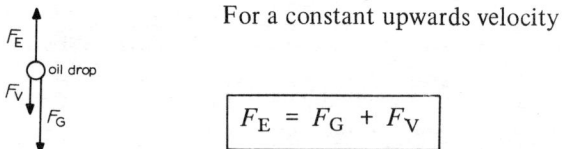

For a constant upwards velocity

$$\boxed{F_E = F_G + F_V}$$

E6

(a) The parallel section must have an effective resistance of 2 Ω, since $2 \text{ }\Omega + 8 \text{ }\Omega = 10 \text{ }\Omega$.

Using $\dfrac{1}{R_p} = \dfrac{1}{6} + \dfrac{1}{X}$, we get

$\dfrac{1}{X} = \dfrac{1}{2} - \dfrac{1}{6} \Rightarrow X = 3 \text{ }\Omega$

(b) Applying the parallel combination formula, we get

$$\frac{1}{4} = \frac{1}{8} + \frac{1}{(6+Y)}$$

leading to

$Y = 2 \,\Omega$

(c) *Circuit AB* : Current from supply is $I_S = \dfrac{V}{R_{AB}} = \dfrac{60}{10} = 6$ A.

This current branches, with 2 A flowing through the 6 Ω resistor and 4 A flowing through the 3 Ω resistor (X).

Circuit CD : Current from supply is $I_S = \dfrac{V}{R_{CD}} = \dfrac{60}{4} = 15$ A.

This current branches, 7.5 A flowing through top and bottom branches because each branch has the same resistance (8 Ω).

E7

(a) (i) *Voltmeter* : With switches open, the current must flow in path T_1 A B C T_2 through the coil and the series multiplier resistor R_2.
(ii) *Ammeter* : With switches closed, R_2 has been shorted out, whereas R_1 becomes a shunt resistor in parallel with meter M. There are two routes for current, through meter T_1 A B E F C T_2, and through shunt T_1 A D E F C T_2.

(b) *Voltmeter* 0 - 1 V

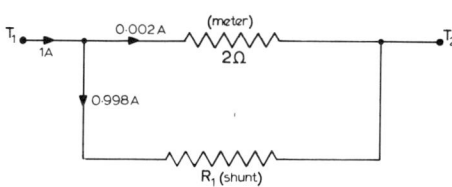

p.d. across T_1 T_2 must be 1 V when 0.002 A flows, therefore resistance between

T_1 and T_2 = $R_{TOTAL} = \dfrac{V}{I} = \dfrac{1 \text{ V}}{0.002 \text{ A}} = 500 \,\Omega$.

$R_{TOTAL} = R_{METER} + R_{MULTIPLIER}$ $\therefore$ 500 Ω = 2 Ω + 498 Ω.

A multiplier of resistance 498 Ω (R_2) is required.

Ammeter 0 - 1 A

When a current of 1 A flows into the ammeter, 0.002 A flows through the coil $\Rightarrow$ 0.998 A is shunted through R_1. Since shunt and meter are in parallel

p.d. across shunt = p.d. across motor

$0.998 \times R_1 = 0.002 \times 2$ $\therefore R_1 = 0.004 \,\Omega$ (approximately)

(c) $0.5 \times 0.998 = 0.499$ A
(d) $0.5 \times 0.002 = 0.001$ A

64   Electricity

To summarize :
Voltmeter Conversion : Connect a high resistance multiplier in series.
Ammeter Conversion : Connect a low resistance shunt in parallel.

E8
(a) The e.m.f. is the voltmeter reading on open circuit therefore e.m.f. = 2.0 V.
(b) The t.p.d. is the voltmeter reading when circuit is in action therefore t.p.d. = 1.6 V
(c) Applying Ohm's Law to the external part of the circuit, we get

$$R_{EXT} = \frac{V_{EXT}}{I} = \frac{t.p.d.}{I} = \frac{1.6 \text{ V}}{0.8 \text{ A}} = 2.0 \text{ } \Omega$$

Applying Ohm's Law to the internal part of the circuit, we get

$$R_{INT} = \frac{V_{INT}}{I} = \frac{(e.m.f. - t.p.d.)}{I} = \frac{(2.0 \text{ V} - 1.6 \text{ V})}{0.8 \text{ A}} = 0.5 \text{ } \Omega$$

(d) The new value of $R_{EXT} = \frac{1}{2} \times 2.0 \text{ } \Omega = 1.0 \text{ } \Omega$.

$R_{TOTAL} = R_{EXT} + R_{INT} = 1.0 \text{ } \Omega + 0.5 \text{ } \Omega = 1.5 \text{ } \Omega$

Now that we know the total resistance of the entire circuit, we can get the current as follows :

$$I = \frac{V_{TOTAL}}{R_{TOTAL}} = \frac{e.m.f.}{R_{TOTAL}} = \frac{2.0 \text{ V}}{1.5 \text{ } \Omega} = 1.3(3) \text{ A}$$

which is the ammeter reading.
Since the voltmeter is recording the t.p.d. (V_{EXT}), we can now find its new reading as follows :

$V_{EXT} = I R_{EXT} = 1.3(3) \times 1.0 = 1.3(3)$ V.

(e) The term 'lost volts' refers to the difference between the e.m.f. and the t.p.d. It is a measure of the energy wasted per coulomb in driving charge through the cell against the internal resistance.

E9
(a) When the bulb is connected, current flows through **two resistances in series** :
  (i) the external resistance of the circuit due to the bulb (R_{EXT}) and
  (ii) the internal resistance of the cell (R_{INT}).
Therefore a p.d. must develop across the bulb (V_{EXT}) and a p.d. must develop within the cell (V_{INT}). Obviously

e.m.f. = $V_{EXT} + V_{INT}$

Now the voltmeter in Fig. II is across the external part of the circuit and so it measures V_{EXT}, which must be less than the e.m.f. of the cell because some p.d. has developed within the cell. In this example

e.m.f. = 1.5 V, V_{EXT} = 1.2 V, so V_{INT} = 0.3 V.

(b) Since the *same* current flows through R_{EXT} and R_{INT}, the values of V_{EXT} and V_{INT} can be found from Ohm's Law.

$$V_{EXT} = I R_{EXT} \quad \text{and} \quad V_{INT} = I R_{INT} \Rightarrow \frac{V_{EXT}}{V_{INT}} = \frac{R_{EXT}}{R_{INT}}$$

If a lower internal resistance is used then the ratio $\dfrac{R_{EXT}}{R_{INT}}$ increases

and so does the ratio $\dfrac{V_{EXT}}{V_{INT}}$. This means that the external voltage will increase at the expense of the internal voltage, making the bulb brighter.

(c) (i) $R_{EXT} = \dfrac{V_{EXT}}{I} = \dfrac{1.2\ \text{V}}{0.3\ \text{A}} = 4\ \Omega$

(ii) $R_{INT} = \dfrac{V_{INT}}{I} = \dfrac{(1.5 - 1.2)\ \text{V}}{0.3\ \text{A}} = 1\ \Omega$

E10
(a) *Tom's circuit* would be a disaster because the sensitive galvanometer G_1 has been connected straight across the driver cell. Far too large a current would flow through G_1 when the switch was closed, almost certainly wrecking it.
(b) *Dick's circuit* is a balanced Wheatstone Bridge because the ratio of resistances in the top branch is equal to the ratio of resistances in the bottom branch.

$$\dfrac{9\ \Omega}{12\ \Omega} = \dfrac{12\ \Omega}{16\ \Omega}$$

At balance there is no current in G_2 when the switch is closed.
(c) *Barry's circuit* is an off-balance Wheatstone Bridge because the ratio of resistances in the top branch is *not* the same as the ratio of resistances in the bottom branch.

$$\dfrac{5\ \Omega}{10\ \Omega} \neq \dfrac{10\ \Omega}{5\ \Omega}$$

If we calculate the p.d.s. across the resistors in Barry's Circuit, we get

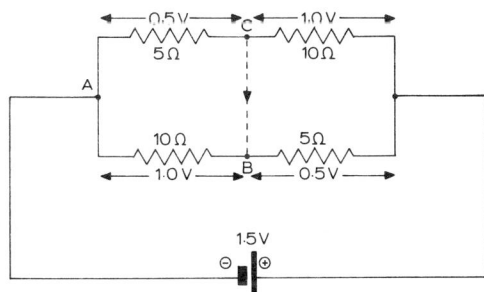

We see that, relative to point A, point B is at a more +ve potential than point C. Now electrons flow towards more +ve potentials and so the electron current is directed through G_3 towards B.

E11
(a) The galvanometer must be sensitive to detect the small currents which flow through it when the bridge is only slightly off the balance position. The galvanometer must be centre zero, because currents will flow one way or the other through it if the slider T is positioned to the left or right of the balance point.

66   Electricity

(b) Switch S_1 brings the driver cell into the circuit. It should be closed only when measurements are being made, otherwise the cell will be 'drained' unneccessarily and the wire will become overheated.
Switch S_2 is part of a sensitivity control. When open, the galvanometer current is kept low by the protective resistor R_p, preventing overloading the galvanometer when the bridge is off-balance. When S_2 is closed, R_p is shorted out and the galvanometer is unprotected. This is done when the bridge is known to be very nearly balanced, to produce maximum sensitivity.

(c) For a balanced bridge,

$$\frac{6.00}{R_U} = \frac{AT}{TB} = \frac{40 \text{ cm}}{60 \text{ cm}} \Rightarrow R_U = 9.00 \ \Omega$$

(d) (i) The balance position would not be altered at all.
(ii) The effective resistance in the top left section of the bridge would now be $3.00 \ \Omega$ (the combination of two $6.00 \ \Omega$ resistors in parallel).
For a balanced bridge,

$$\frac{3.00}{9.00} = \frac{AT'}{T'B}$$

which gives $AT' = 25$ cm.

E12

(a) potential gradient = $\dfrac{\text{potential difference across a section of wire}}{\text{length of the section of wire}}$

To obtain an accurate value, consider section AT :

potential gradient = $\dfrac{V_{AT}}{AT} = \dfrac{1.10 \text{ V}}{0.44 \text{ m}} = 2.50 \text{ V m}^{-1}$

(b) ratio of e.m.f.s. = ratio of balance lengths

$$\Rightarrow \frac{E_U}{E_S} = \frac{AT_U}{AT_S} \Rightarrow E_U = \frac{E_S \times AT_U}{AT_S} = \frac{1.10 \times 0.80}{0.44} = 2.0 \text{ V}$$

(c) p.d. across AB = 2.50 V, because AB = 1.00 m.
resistance of AB = 25 Ω, because AB = 1.00 m.

$\therefore$ current through wire AB = $I_{AB} = \dfrac{V_{AB}}{R_{AB}} = \dfrac{2.50 \text{ V}}{25 \ \Omega} = 0.10$ A.

(d) (i) If the wire has a uniform cross-section, its resistance per cm is constant and so the potential gradient along it is constant.
(ii) If the resistance per metre were low, a large current would flow through AB making it heat up, expand and distort.

(e) R_p is a resistor which protects G by keeping the value of the current down to a safe value while the balance position is being located.
Once the approximate balance position has been found, switch K is closed to short out R_p. This improves the sensitivity of the measurement.

E13

(a) When the rheostat slider is at **p**, the entire voltage of the driver cell is being applied to the wire AB therefore $V_{AB} = E_d = 2.0$ V.

$$\frac{E_S}{E_d} = \frac{AT}{AB} \Rightarrow \frac{1.20 \text{ V}}{2.0 \text{ V}} = \frac{AT}{5.00} \quad \therefore AT = 3.00 \text{ m}$$

When the slider is at q, the driver cell voltage is being applied across a 20 Ω rheostat and a 60 Ω (12 x 5.00) wire in series. The reduced value of the p.d. across the wire is given by

$$V_{AB} = \left(\frac{60}{60+20}\right) \times 2.0 \text{ V} = 1.5 \text{ V}$$

Since e.m.f. ratio = balance length ratio, we see that

$$\frac{E_S}{V_{AB}} = \frac{AT}{AB} \Rightarrow \frac{1.20 \text{ V}}{1.50 \text{ V}} = \frac{AT}{5.00} \quad \therefore AT = 4.00 \text{ m}$$

(b) The e.m.f. of the standard cell is 1.20 V. When the 20 Ω rheostat is placed across this cell, the t.p.d. drops due to internal resistance effects. In fact,

$$\frac{E_S}{V_{EXT}} = \frac{3.00}{2.50} \Rightarrow \frac{1.20 \text{ V}}{V_{EXT}} = \frac{3.00}{2.50} \quad \therefore V_{EXT} = 1.00 \text{ V}$$

The t.p.d. of the cell is 1.00 V. The internal resistance is

$$R_{INT} = \frac{V_{INT}}{I} = \frac{(1.20 - 1.00) \text{ V}}{\left(\frac{1.00 \text{ V}}{20 \text{ Ω}}\right)} = 4.0 \text{ Ω}$$

E14

(a) (i) The electromagnet acts like a bar magnet with a south pole at the end nearer the speech coil. The speech coil acts like a bar magnet whose poles are reversing every time the alternating current reverses. The speech coil is attracted to and then repelled from the electromagnet and the vibration of the speech coil is transmitted to the cone.

(ii) Use the left-hand rule to identify the poles of the two coils.

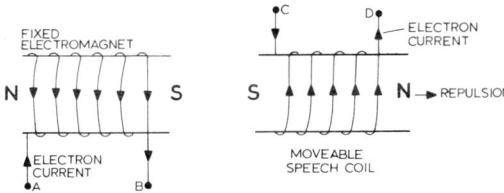

Obviously the south poles repel so the speech coil moves out.

(b) (i) A wire carrying current in a magnetic field is acted upon by a force at right angles to the field and the wire (Right-hand Motor Rule). As the current is alternating at 50 Hz, the motor force on the wire is reversing every $\frac{1}{100}$ s and so a standing wave pattern can be produced, provided the tension and the length of the wire are correctly matched.

(ii) There are five full waves (11 nodes) between A and D

$$\Rightarrow \lambda = \frac{AB}{5} = 12 \text{ cm}$$

Also f = 50 Hz. From the wave equation, we get

$$v = f\lambda = 50 \times 0.12 = 6.0 \text{ m s}^{-1}$$

E15

(a) (i) The reason why the current through the coil takes longer to reach its final value is connected with the magnetic properties of a coil (inductor). When a current is first passed through a coil, a magnetic field grows around the coil. Since moving magnetic fields induce back e.m.f.s, according to the laws of electromagnetic induction, an e.m.f. is created across the coil which opposes the growing current and tends to keep its value down. It is this back e.m.f. which produces the observed delay in bulb B_2 reaching full brightness. The resistances of the coil and the resistor must be the same as we are told that the final brightnesses of the bulbs are the same.

(ii) The current through bulb B_1 would decrease due to the larger resistance. The current through bulb B_2 would rise more quickly to the final value because the inductance of the coil has been reduced by the removal of its iron core.

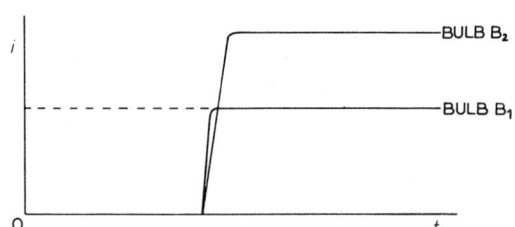

(b) Transformers must never be used with d.c. supplies because the resistance of the primary winding is usually quite low and a large d.c. current would flow through the primary, probably burning it out. With a.c., there is always a back e.m.f. associated with the changing magnetic field and this keeps the current down to a safe value. Also, continually changing primary currents are needed to induce e.m.f.s in the secondary.

E16

(a) turns ratio = voltage ratio

$$\frac{T_p}{T_s} = \frac{V_p}{V_s} = \frac{240}{12} = 20$$

(b) primary power P_p = secondary power P_s

$V_p I_p = V_s I_s$

(i) Only one bulb is lit $\therefore P_s = 24$ W, $V_s = 12$ V.

Secondary current is $I_s = \dfrac{P_s}{V_s} = \dfrac{24 \text{ W}}{12 \text{ V}} = 2$ A.

Also the primary current I_p is obtained from $240 I_p = 12 \times 2$

$\Rightarrow I_p = 0.1$ A.

(ii) Now there are three bulbs therefore $P_s = 3 \times 24 = 72$ W, $V_s = 12$ V.

Secondary current is $I_s = \dfrac{P_s}{V_s} = \dfrac{72 \text{ W}}{12 \text{ V}} = 6$ A.

Also the primary current I_p is obtained from $240 I_p = 12 \times 6 \Rightarrow I_p = 0.3$ A.

(c) (i) p.d. across R = (230 V – 115 V) = 115 V.
current through R = 5 A (resistor is in series with projector).
$$R = \frac{V}{I} = \frac{115 \text{ V}}{5 \text{ A}} = 23 \text{ }\Omega$$
(ii) $\frac{N_p}{N_s} = \frac{960}{N_s}$, and $\frac{V_p}{V_s} = \frac{230 \text{ V}}{115 \text{ V}} = 2 \Rightarrow \frac{960}{N_s} = 2 \therefore N_s = 480$ turns.

(iii) *Circuit I*
$$\% \text{ efficiency} = \frac{\text{useful power output}}{\text{total power input}} \times 100 = \frac{(115 \times 5) \times 100}{(230 \times 5)} = 50\%$$

Circuit II Most commerical transformers have a high % efficiency, say 90%.
(iv) I_x = 5 A, I_y = 5A, I_z = 2.5 A $\therefore I_z$ is smallest.

E17

(a) current = $\frac{\text{power}}{\text{voltage}}$, or $I = \frac{P}{V}$. This expression is applied as follows :

(i) $I = \frac{600 \text{ kW}}{4 \text{ kV}}$ = 150 A

(ii) $I = \frac{600 \text{ kW}}{400 \text{ kV}}$ = 1.5 A

(iii) For the step-down transformer, which is 100% efficient.
$$V_p I_p = V_s I_s \Rightarrow I_s = \frac{V_p I_p}{V_s} = \frac{391\,000 \times 1.5}{230} = 2550 \text{ A}$$

(b) input voltage to cables = 400 kV, and output voltage from cables = 391 kV.
Thus the potential drop along the actual cables is (400 – 391) kV = 9 kV.
The total cable resistance is now found from Ohm's Law, as follows :

$$R_\text{cable} = \frac{V_\text{cable}}{I_\text{cable}} = \frac{9000 \text{ V}}{1.5 \text{ A}} = 6000 \text{ }\Omega$$

The total length of cable is 2 × 150 = 300 km. Thus the resistance of the cable per km is $\frac{6000 \text{ }\Omega}{300 \text{ km}} = 20 \text{ }\Omega \text{ km}^{-1}$.

(c) step-down ratio = voltage ratio = $\frac{391\,000 \text{ V}}{230 \text{ V}}$ = 1700 : 1

(d) To transmit the power from generator to consumer load, a current has to flow along the cables. Power is therefore wasted in the form of heat losses in the cables. This cable power loss is found using $P = I^2 R$, where I is the current in the cable and R is the total resistance of the cable. Obviously, for a given length and construction of cable, the power waste depends upon the value of I^2. To keep this small, we must make I as small as we can. This can only be done by using a step-up transformer to transmit the power at **low current : high voltage**. The method the economics student suggests would transmit the power at **high current : low voltage**, leading to a colossal loss of power in the form of heat in the cables.

E18

(a) (i) increase in k.e. = work done on the electron
= force on electron × distance moved by electron
$$= \left(\frac{qV}{d}\right) \times d = qV \text{ (as required).}$$

(ii) To obtain increase in speed, we use $\frac{1}{2}mv^2 = qV$ (see (i)). Solving for v, we get

$$v = \sqrt{\frac{2qV}{m}}$$

(iii) Substitution:

$$v = \sqrt{\frac{2 \times 1.6 \times 10^{-19} \times 4.5 \times 10^3}{9.0 \times 10^{-31}}} = \sqrt{1.6 \times 10^{15}} = \sqrt{16 \times 10^{14}}$$

$\therefore v = 4.0 \times 10^7$ m s^{-1}.

(b) If N electrons move across the tube in a second, the total charge transferred must be $N \times q$. But current is the charge flowing per second $\therefore I = Nq$.

Thus $N = \dfrac{I}{q} = \dfrac{2.4 \times 10^{-6} \text{ A}}{1.6 \times 10^{-19} \text{ C}} = 1.5 \times 10^{13}$ electrons each second.

E19

(a) When S is in contact with x, the diode conducts and a current flows round the circuit through the two series resistors. The p.d. developed across the 3 Ω resistor is recorded on the trace at A, B and C. The height of the trace at these positions is determined by the size of the p.d. across the 3 Ω resistor, which is in fact 0.6 V ($\frac{3}{5} \times 1.0$ V).
While the contact is moving between x and y, there is obviously no deflection on the trace. Likewise there is no deflection while the contact is at y, because the diode is reverse-biased (look carefully at the lower cell).

(b) The distance between the starts of pulses A and B, measured horizontally, is 4 cm and so represents a time interval of 40 ms (4 cm × 10 ms cm^{-1}). The frequency of the switch is $\therefore$ given by

$$f = \frac{1}{T} = \frac{1}{0.040 \text{ s}} = 25 \text{ Hz}$$

(c) The diode would now conduct in switch position y only. The current would flow round the circuit in the opposite direction, reversing the direction of the p.d. across the 3 Ω resistor. Thus the trace would be like this:

(d) The height of the deflection would be less, because the p.d. across the 2 Ω resistor is only 0.4 V, instead of 0.6 V.

(e) The trace would show the typical sine wave outline, but would be half-wave rectified like this:

E20
(a)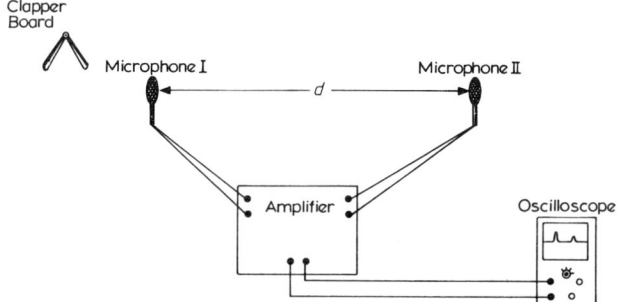

When the clapper board is used, a sound pulse travels past the two microphones. Two small electrical pulses are produced in the microphones and amplified before being fed to the Y-plates of the oscilloscope. Blips are 4 cm apart on the screen and so the time interval between them is 4 cm × 10 ms cm^{-1}, or t = 40 ms = 0.040 s. If the distance d between the microphones is 13.2 m, then the velocity of sound is

$$v = \frac{d}{t} = \frac{13.2 \text{ m}}{0.040 \text{ m}} = 330 \text{ m s}^{-1}$$

(b) There are ten full waves showing on the screen and so each occupies 1 cm. Therefore 1 full wave is produced in 10 ms ($\frac{1}{100}$ s).
It follows that the frequency must be 100 Hz.

$$\text{frequency of signal}, f = \frac{1}{T} = \frac{1}{0.01 \text{ s}} = 100 \text{ Hz}$$

(c) Amplitude increases but number of waves stays the same.
(d) (i) At 1 ms cm^{-1}, instead of 10 ms cm^{-1}, the spot is travelling across the screen ten times faster. The new TRACE A would be confusing because the spot would make several scans of the screen in the time between the blips.
(ii) We would obtain 1 full wave, instead of ten, on the new TRACE B.

E21
(a) The supply voltage is applied across a resistance wire AB. By moving the sliding contact T along this uniform wire, any desired fraction of the supply voltage can be developed across AT, and hence across the bulb. With T near A, the p.d. across the bulb is small. With T near B, the p.d. across the bulb is almost the same as the supply voltage.
(b) A photocell could be used to measure the brightness of the bulb. The brightness levels would be the same when the photocell meter readings were the same. Obviously, the photocell would have to be kept at a fixed distance away from the bulb, and it would have to be enclosed to prevent stray light affecting its response.
(c) The 'root mean square' voltage is the d.c. voltage which produces the same heating effect as a given a.c. voltage when it is applied to the same resistor (a bulb in this case).
(d) d.c. voltage across the bulb = 3.6 V (the r.m.s. voltage).
a.c. voltage across the bulb = 5.1 V from zero to peak (the peak voltage).

$$V_{rms} = \frac{3.6}{5.1} \times V_{peak} = 0.71 \times V_{peak}$$

$$V_{peak} = \frac{5.1}{3.6} \times V_{r.m.s.} = 1.42 \times V_{r.m.s.}$$

N.B. in theory, the relationships are :

$$V_{r.m.s.} = \frac{1}{\sqrt{2}} \times V_{peak} \quad \text{and} \quad V_{peak} = \sqrt{2} \times V_{r.m.s.}$$

(e)
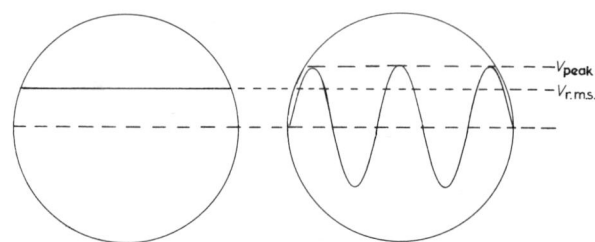

E22
(a) When the current flowing through a coil is *altered*, the associated magnetic field induces a back e.m.f. across the coil. The back e.m.f. partly cancels out the effectiveness of the supply e.m.f. and so the circuit current is reduced.

When an a.c. supply is used, the current is continually changing in size and direction, and so the coil continually has a back e.m.f. developed across it and this acts as a control on the level of the alternating current.

With a soft-iron core, the coil acts like a *stronger* electromagnet and so *larger* back e.m.f.s. are induced by the varying current. The increased opposition (inductive reactance) reduces the size of the circuit current and the bulb gets dimmer.
(b) With a higher frequency setting, the circuit current and the magnetic field round the coil vary more rapidly. By Faraday's Law, the induced e.m.f. is proportional to the rate of change of the magnetic field. Thus at higher frequencies, the back e.m.f. is greater so the circuit current is reduced (the inductive reactance increases) making the bulb dimmer.
(c) Dimmer at 1100 Hz (see above).
(d) The coil is called an **inductor**.
(e) The opposition is called **inductive reactance**.
(f) This keeps the normal resistance of the coil low because (i) copper is a good conductor and (ii) resistance reduces when the cross-sectional area of the wire increases.
(g) $V_{peak} = \sqrt{2} \times V_{r.m.s.} = \sqrt{2} \times 6.0 = 8.5$ V

E23
(a) By examining the pattern of the results, we see that $I_L = 12$ mA at $f = 100$ Hz, and $I_L = 2.4$ mA at $f = 500$ Hz.

(i) $X_L = \dfrac{12.0 \text{ V}}{0.012 \text{ A}} = 1000 \, \Omega$

(ii) $X_L = \dfrac{12.0 \text{ V}}{0.0024 \text{ A}} = 5000 \, \Omega$.

(b) $I_{peak} = \sqrt{2} \times 4.0 = 5.7$ mA.

(c)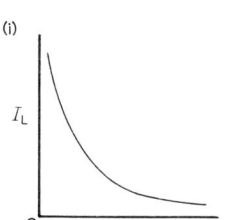

(d) (i) I_L increases because (ii) X_L reduces for any chosen frequency.

E24
(a) The capacitor charges up, discharges, then re-charges in the opposite direction during each cycle of the generator output. Electrons are moving on and off the plates of the capacitor at the frequency of the supply. The charging and discharging currents flow through the bulb, which therefore lights up.
(b) With a higher frequency, a greater total charge per second moves through the bulb, producing more heat per second in the filament. The bulb therefore lights up brighter.
(c) With a larger capacitance, more charge moves on and off the plates during each cycle and so the bulb lights up brighter.
(d) The graph would be the usual sine shape. Since the frequency is set at 1000 Hz, the time for one complete cycle (period) is 0.001 s. Since the r.m.s. voltage is 6.0 V, the peak voltage is $\sqrt{2} \times 6.0$, or 8.5 V.

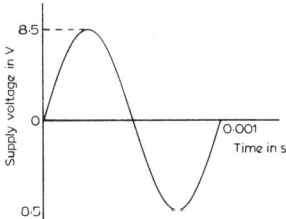

E25
(a) From the pattern of the results given, it is seen that I_C = 16 mA when f = 100 Hz and that I_C = 80 mA when f = 500 Hz.

(i) $X_C = \dfrac{12.0 \text{ V}}{0.016 \text{ A}} = 750\ \Omega$

(ii) $X_C = \dfrac{12.0 \text{ V}}{0.080 \text{ A}} = 150\ \Omega$

(b)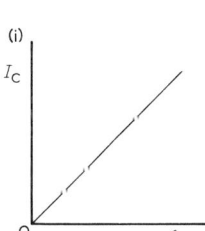

(c) (i) current reduces (ii) reduces (iii) reduces (iv) increases (v) increases

E26

(a) (i) f_r reduces when L is increased.
(ii) f_r reduces when C is increased.
(iii) The value of f_r is not affected by the value of resistance. However, the peak and dip at resonance would be flatter with resistance included. The circuits would become less selective.

(b)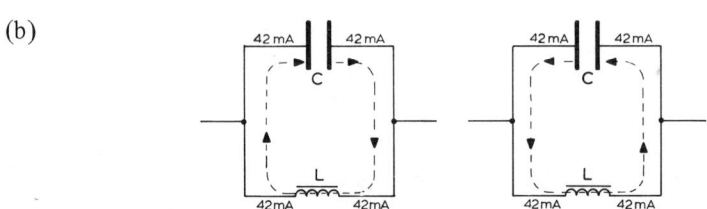

The currents in C and L are *out of phase* so that the current is to the right in the C branch when it is going to the left in the L branch. The overall effect is as if a current of 42 mA were oscillating within the L–C loop. A small current has to be delivered by the supply to compensate for small energy losses which occur in the form of heat, mainly in the inductor.

(c) Used in the tuning part of radio receivers to select a particular broadcast frequency from the large number of frequencies passing over the aerial.

E27

(a) Circuit A includes an inductor and a milliammeter in series.
Circuit B includes a capacitor and a milliammeter in series.
Circuit C includes a capacitor and inductor in parallel, with the milliammeter in the supply lead.
Circuit D includes a capacitor and inductor in series with a milliammeter.
In all cases the supply should be a signal generator.

(b)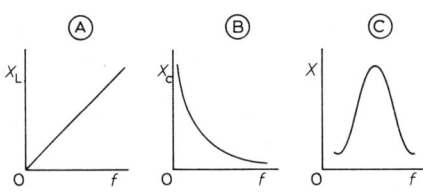

Practice Questions

1. (E1, 2)

Two metal plates, each of area 1.0 m², are held 1 cm apart, forming a simple capacitor. The plates of the capacitor are charged by connecting them across the terminals of a 100 V battery.

(a) How would the *charge stored* be affected by each of the following changes, considered separately?
   (i) Using a 200 V battery instead.
   (ii) Holding the plates 2 cm apart.
   (iii) Reducing the area of overlap by shifting the plates sideways.
   (iv) Inserting a sheet of polythene (dielectric) between the plates.
(b) How would the **capacitance** of the system be affected by each of the changes mentioned in (a)?

2. (E2)

(a) The capacitance (C) of a parallel-plate capacitor depends upon the area of overlap of the plates (A), the separation of the plates (d) and the dielectric constant of the material between the plates (k). Write down the correct relationship for the capacitance in terms of A, d and k.

(b) Which of the following equations represents the correct definition of capacitance?

(i) $C = QV$ (ii) $C = \dfrac{V}{Q}$ (iii) $C = \dfrac{Q}{V}$

(c) What are the correct SI units for C, Q and V?

(d) What do the following prefixes stand for?
(i) milli (ii) micro (iii) nano (iv) pico

3. (E2)

(a) A capacitor stores 2.4×10^{-5} coulombs of charge when the p.d. across its plates is 6.0 V. Calculate its capacitance in microfarads.

(b) A 0.1 μF capacitor is connected across a 12 V battery. How much charge is stored?

(c) Why is a maximum permitted voltage often stamped on the case of a capacitor?

4. (E3)

To control the rate at which an 8 μF capacitor charges up from a 4 V battery, a pupil inserts a 1 MΩ resistor in series with the capacitor.

(a) What is the initial p.d. across (i) the capacitor? (ii) the resistor?

(b) What is the initial current through the resistor?

(c) What is the final p.d. across (i) the capacitor? (ii) the resistor?

(d) What is the final current through the resistor?

(e) Draw a sketch graph to show how the charging current changes with time while the capacitor is charging up.

5. (E3)

(a) Referring to the charging circuit described in question 4 (E3), how would the charging rate be affected by each of the following changes, taken independently?
   (i) Using a 16 μF capacitor instead of the 8 μF capacitor.
   (ii) Using an 8 V battery instead of a 4 V battery.
   (iii) Using a 2 MΩ resistor instead of a 1MΩ resistor.

(b) The 'half-life' of a particular RC charging circuit is 9.0 s and the initial charging current is 80 μA.
   (i) What time interval would pass before the charging current reduced to 5 μA?
   (ii) What would be the value of the charging current after 27 s?
   (iii) What could be done to increase the 'half-life' to 18 s?

6. (E4)

A 2.0 V cell is connected across a 2.0 μF capacitor in series with a 2.0 MΩ resistor.

(a) Draw two graphs to indicate how the p.d. across (i) the capacitor and (ii) the resistor varies with time during the charging.

(b) What is the final p.d. across the capacitor and the final charge stored by it?

7. (E4)

Referring to the components in question 6 (E4), the capacitor is disconnected when it is fully charged. It is then allowed to discharge by fitting the 2.0 MΩ resistor directly across the terminals of the capacitor.

76 Electricity

(a) Draw two graphs to indicate how the p.d. across (i) the capacitor (ii) the resistor varies in time during the discharging.
(b) What is the final charge on and p.d. across the capacitor?

8. (E5)
In a Millikan oil drop experiment, two conducting plates are positioned 1.6 cm apart. When the p.d. across the plates is raised to 360 V, an oil drop of mass 7.2×10^{-16} kg is suspended at rest mid-way between the plates. The oil drop is known to be negatively charged.
(a) Calculate the charge carried by the oil drop.
(b) How many excess electrons does the oil drop carry?
(c) The p.d. across the plates is now increased slightly. Describe the motion of the oil drop.
(d) When a charged oil drop falls between the two *uncharged* conducting plates, the drop quickly reaches a terminal velocity. What forces are acting upon the oil drop when such a motion is obtained?

9. (E6)
You are given three resistors of values 3 Ω, 6 Ω and 12 Ω. Using all three resistors, how many different arrangements can you make? State the effective resistance of each arrangement.

10. (E6)
Three resistors of values 1 Ω, 2 Ω and 3 Ω are connected across a 12 V d.c. supply, first in series, and then in parallel. Fill in the current values (a) to (f), the p.d. values (g) to (l), and the power values (m) to (r).

	Series			Parallel		
Resistor	1 Ω	2 Ω	3 Ω	1 Ω	2 Ω	3 Ω
Current through resistor	(a)	(b)	(c)	(d)	(e)	(f)
p.d. across resistor	(g)	(h)	(i)	(j)	(k)	(l)
Power developed in resistor	(m)	(n)	(o)	(p)	(q)	(r)

11. (E7)
The coil resistance of a galvanometer is 5 Ω and it gives a full scale deflection when a current of 10 mA passes through the coil.
(a) How would you convert it into a voltmeter whose range is 0-100 V?
(b) How would you convert it into an ammeter whose range is 0-10 A?

12. (E7)
Two 1000 Ω resistors are connected in series across a constant d.c. supply of value 12.0 V.
(a) What would be recorded on an *ideal* voltmeter placed across one of the resistors?
(b) What would be recorded on a voltmeter of resistance 1000 Ω when placed across one of the resistors?
(c) Would the difference between the readings of (a) and (b) be increased or decreased by using two 100 Ω resistors in series instead?

Practice questions 77

13. (E8)
A pupil is given a very good (very high resistance) voltmeter and a torch battery. When he fits the voltmeter across the terminals of the battery it registers 4.5 V but when he connects the battery across a 6.0 Ω resistor, the voltmeter reading across the battery terminals decreases to 3.0 V.
(a) Calculate the internal resistance of the battery.
(b) What value of resistor would have to be connected across the battery to reduce the voltage reading to 2.5 V?

14. (E9)
When the terminals of a 12.0 V car battery are shorted a current of 100 A flows through the 'short'. What is the internal resistance of the car battery?

15. (E9)
A battery of e.m.f. 12.0 V and internal resistance 1.0 Ω is connected across a rheostat. By moving the slider of the rheostat, the current taken from the battery is progressively increased up to 10.0 A.
(a) Draw an accurate graph, including numerical values, to show how the t.p.d. of the battery is related to the current it supplies.
(b) What is the resistance of the rheostat when it is set to give I = 10.0 A?

16. (E10)
A Wheatstone Bridge consists of three resistors of known resistances and one resistor whose resistance is required.
(a) What is the relationship amongst the four resistors at balance?
(b) What instrument is used to indicate when balance has been achieved?
(c) What characteristics should this detecting instrument have?
(d) Is the balance position of the Wheatstone Bridge affected by
  (i) changing the e.m.f. of the driver cell?
  (ii) reversing the terminals of the driver cell?

17. (E11)
Study the metre-bridge circuit shown in problem E11. The length of the resistance wire AB is 100 cm. The value of the standard resistor is 6.00 Ω.
(a) What is the value of the unknown resistor R_U if TB = 70 cm?
(b) What would be the balance length AT if R_U were replaced by an 18.00 Ω resistor?

18. (E12)
A potentiometer circuit like the one shown in problem E12 has a uniform resistance wire AB which is 5.00 m long. When a standard cell of e.m.f. 1.10 V (E_S) is used, the galvanometer registers zero with AT = 2.20 m. When another cell is used instead of the standard cell, the new balance position is AT = 3.80 m.
(a) Calculate the potential gradient along wire AB.
(b) What is the e.m.f. (E_U) of the replacement cell?
(c) Estimate the e.m.f. of the driver cell (E_d).
(d) The resistance of wire AB is 6.0 Ω m^{-1}. What current flows through the wire while the circuit is balanced?
(e) How would balance lengths be affected if a resistor were fitted in series with the driver cell?

78   Electricity

19. (E16)
(a) A transformer is used to reduce an alternating voltage from 240 V to 6 V. An immersion heater rated '6 V, 36 W' is connected across the secondary and switched on for 5 minutes. Assuming that the transformer is 100% efficient, calculate
   (i) the number of turns on the secondary if there are 960 turns on the primary.
   (ii) the secondary current.
   (iii) the primary current.
(b) A step-up transformer, which is 90% efficient, is used to increase an alternating voltage from 240 V to 360 V. If the secondary delivers 10 mA at 360 V to the next stage of a power supply, what is the primary current?

20. (E17)
The total length of grid cables connecting the generator to the consumer in a national grid system is 200 km and the cable has a resistance of 4 Ω km^{-1}. (For a similar diagram, see E17). The cable is carrying a current of 5 A.
(a) How much power is wasted as heat in the cable?
(b) The input voltage to the cables is 120 kV. What is the output voltage from the cables?
(c) What is the turns ratio of the step-down transformer if the consumer is supplied at 240 V?
(d) What is the turns ratio of the step-up transformer if the generator operates at 4 kV?

21. (E18)
When a p.d. of 180 V is applied across the electrodes of a thermionic diode, a current (I) of 8.0 mA flows in the anode lead.
(a) Estimate the speed (v) at which the electrons strike the anode after passing through the diode.
(b) Estimate how many electrons (N) hit the anode per second.
(c) If the area of the anode is A, show that the average pressure exerted by the electron beam upon the anode is given by the expression

$$p = \frac{mvI}{Aq}$$

where q is the electronic charge, and m is the mass of the electron.
(Electronic charge = 1.6×10^{-19} C; Mass of electron = 9.0×10^{-31} kg)

22. (E19)
A signal generator operating at 100 Hz is connected across a diode and a resistor in series. The p.d. across the resistor is displayed on an oscilloscope whose screen measures 10 cm x 10 cm, and whose time-base is set at '5 ms/cm'.
(a) How long does the spot take to make one scan across the oscilloscope screen?
(b) Draw the trace you would expect to appear on the screen.
(c) What name is used to describe this waveform?
(d) If the generator frequency is increased to 1000 Hz, what should be done to the time base setting to obtain the same pattern as in (b)?
(e) If the output voltage of the generator is increased, what will be the effect upon the pattern produced on the screen?

23. (E21)
(a) A resistor is connected across a 12 V d.c. battery for 5 minutes. The total energy produced, mainly in the form of heat, is 100 000 J. The same resistor is then connected across a 12 V r.m.s. transformer for 5 minutes. What total energy will be produced?
(b) What is the peak output voltage of a 240 V standard mains supply?

24. (E21)
(a) The root mean square voltage provided by a sine wave generator is 10 V r.m.s. Calculate
   (i) the peak output voltage.
   (ii) the peak-to-peak output voltage.
(b) The above voltage is examined on an oscilloscope whose Y-gain setting is 5 V cm^{-1}. Describe what you would expect to see on the screen while the time-base is in the 'off' position.

25. (E21)
(a) The peak output voltage of a transformer is 8.0 V. What is the r.m.s. output voltage?
(b) When the Y-gain control of an oscilloscope is set at 50 V cm^{-1} and an a.c. input is examined, a vertical line 8.0 cm long is obtained on the screen. Calculate (i) the peak and (ii) the r.m.s. voltage of the a.c. input.

26. (F22)
(a) A coil (inductor) is made up by wrapping 1000 turns of thick copper wire around a hollow plastic tube. This inductor is then connected in series with a bulb across a signal generator whose controls are set at 4.0 V r.m.s., 100 Hz. The bulb lights up to its normal brightness. Using the terms 'dim', 'normal' and 'bright' to describe the light intensity of the bulb, complete the following chart :

Signal generator settings	Light intensity of bulb
4.0 V r.m.s., 100 Hz	normal
4.0 V r.m.s., 80 Hz	?
4.0 V r.m.s., 120 Hz	?
3.0 V r.m.s., 100 Hz	?
5.0 V r.m.s., 100 Hz	?

(b) What would be the effect of introducing an iron core into the hollow tube of the inductor?
(c) What is meant by the term 'inductive reactance'? Suggest a suitable SI unit for this.
(d) Describe how you would construct an inductor with a large value of inductance.

27. (E23)
(a) An inductor is connected across a 10 V r.m.s. supply and the frequency of the supply is increased slowly from 50 Hz to 500 Hz.
   (i) What happens to the alternating current in the inductor?
   (ii) Draw a sketch graph of current against frequency.
(b) At 50 Hz, the r.m.s. current in the inductor is 100 mA.
   (i) What is the inductive reactance at 50 Hz?
   (ii) What is the inductive reactance at 500 Hz?

(c) Complete the following table of results for an inductor :

f	50 Hz	100 Hz	150 Hz
$i_{r.m.s.}$	100 mA	?	?

28. (E24)

(a) A bulb and a 1000 μF capacitor are placed in series across a signal generator whose output is set at 6.0 V r.m.s., 100 Hz. The bulb lights up to its normal brightness. Using the terms 'dim', 'normal' and 'bright' to describe the light intensity of the bulb, complete the following chart :

Signal generator settings	Light intensity of bulb
6.0 V r.m.s., 100 Hz	normal
6.0 V r.m.s., 80 Hz	?
6.0 V r.m.s., 120 Hz	?
5.0 V r.m.s., 100 Hz	?
7.0 V r.m.s., 100 Hz	?

(b) What would be the effect of introducing a capacitor of value 900 μF instead of the 1000 μF one?

(c) What is meant by the term 'capacitive reactance'? Suggest a suitable SI unit for this

29. (E25)

(a) A capacitor is connected across a 10 V r.m.s. supply and the supply frequency is slowly increased from 50 Hz to 500 Hz.
   (i) What happens to the alternating current in this circuit?
   (ii) Draw a sketch graph of current against frequency.
(b) At 50 Hz, the r.m.s. current in this circuit is 100 mA.
   (i) What is the capacitive reactance of the capacitor at 50 Hz?
   (ii) What is the capacitive reactance of the capacitor at 500 Hz?
(c) Draw a sketch graph to show how the capacitive reactance varies with frequency.
(d) Complete the following table of results for the capacitor :

f	50 Hz	100 Hz	150 Hz
$i_{r.m.s.}$	100 mA	?	?

30. (E26)

(a) A resonant circuit is produced by connecting an inductor and capacitor **in series** across a variable frequency supply. At resonance, the circuit current is found to be 100 mA r.m.s. What happens to the value of the circuit current when the frequency is
   (i) reduced below resonance?
   (ii) increased above resonance?

(b) Another resonant circuit is then produced by connecting the same components **in parallel** across a variable frequency supply. At resonance, the current in the supply lead is 5 mA r.m.s. What happens to the value of the circuit current when the frequency is
   (i) reduced below resonance?
   (ii) increased above resonance?
(c) Compare the sizes of the inductor and capacitor currents when the circuit in (b) is (i) below resonance, (ii) at resonance, (iii) above resonance.

31. (E26)
(a) What would happen to the resonant frequency of a **series** L-C circuit if each of the following changes occurred individually?
   (i) Value of inductor increased.
   (ii) Value of capacitor increased.
(b) Answer the same questions, but for a **parallel** L-C resonant circuit.
(c) In fact, the resonant frequency f_r is given by the relationship

$$f_r = \frac{1}{2\pi\sqrt{LC}}$$

What happens to f_r when L and C are both doubled?

UNIT A—Atomic physics

Worked Examples

A1 Half-life
The graphs below show the variation of count-rate of
(i) an alpha source on its own and
(ii) a source made by combining the above alpha source with a beta source.
The count-rates plotted on the graphs have already been corrected for background radiation.

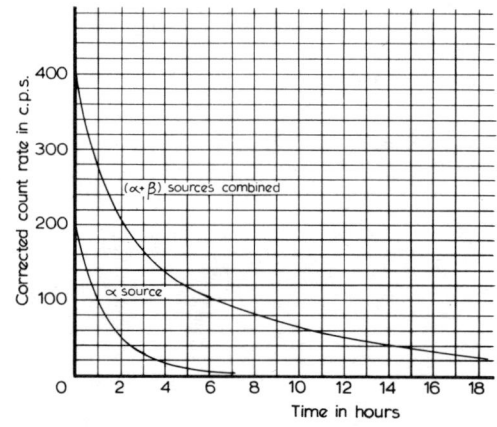

(a) What is the half-life of the alpha source on its own? (3)
(b) What is the half-life of the beta source on its own? (5)
(c) If we try to work out the half-life of the combined source we do *not* get a constant value. However the half-life tends towards a constant value as time progresses. Show that this is true from the graph provided and explain why it is true. (2)

A2 Properties of nuclear radiations
If you complete this table correctly, it could become a useful summary of the properties of nuclear radiations.

Radiation	alpha	beta	gamma
Kind of radiation	particle	(a)	(b)
Speed of radiation	(c)	90% of light	(d)
Charge	+ 2 units	(e)	(f)
Mass	(g)	(h)	zero
Range in air	(i)	(j)	much larger than α & β

Absorbed by	paper	(k)	(1)
Ability to ionise	good ionisers	(m)	(n)
Action in magnetic field	(o)	large deflection	(p)

Copy this chart and complete the blanks (a) to (p), keeping in mind that it is a relative comparison which is required. (10)

A3 Alpha-particle scattering

It was shown by Geiger and Marsden during their experiments on the deflection of alpha particles that alpha particles could be deflected through *large angles* by aiming them at a thin gold foil.

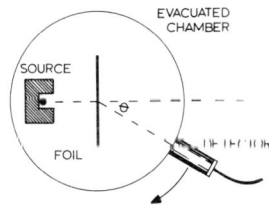

A moveable detector was used to record the number of alpha particles scattered through various angles (shown as θ on the above diagram). The graph below indicates the kind of results obtained :

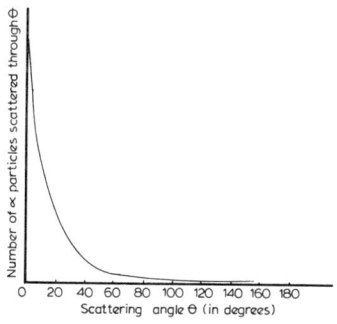

(a) Suggest reasons why the alpha particles should be scattered by the thin gold foil. (2)
(b) Why was it important to evacuate the apparatus? (2)
(c) As a result of this experiment, Rutherford's nuclear model of the atom was developed. Describe this model briefly. (3)
(d) In the graph, you will notice that a small number of alpha particles are scattered through angles greater than 90°.
 (i) What does this suggest about the nuclear collisions involved?
 (ii) Why are the 'back-scattered' particles few in number? (3)

84 Atomic physics

A4 The photoelectric effect

In this diagram, a beam of ultra-violet light from a mercury vapour lamp is falling on a clean zinc plate which is attached to the cap of an electroscope.

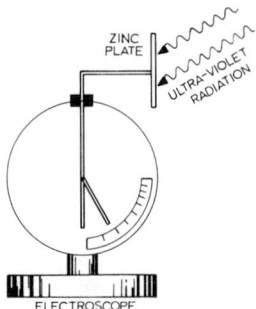

(a) A pupil uses this apparatus to illustrate the photoelectric effect. He starts by charging the electroscope with a negative charge. Explain why it is a combination of (i) electroscope charge, (ii) plate material and (iii) radiation, which enables the photoelectric effect to be shown. (4)

(b) The exposure meter of a camera consists of a photoelectric cell, a battery and a sensitive ammeter which is suitably calibrated for light intensity.

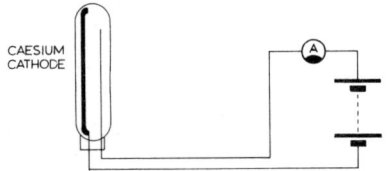

    (i) Why has a caesium photocathode been chosen instead of a zinc one?
    (ii) What factors increase the reading on the ammeter?
    (iii) Colour filters, placed in front of the exposure meter, produce variations in the current reading as shown in Graph I. Increase in light intensity for any particular filter produce variations of current as shown in Graph II.

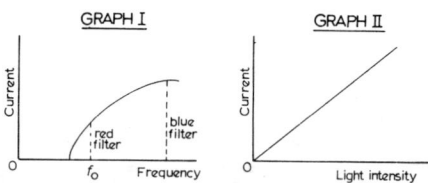

Comment upon the shapes of the two graphs and explain the existence of a threshold frequency f_o in Graph I. (6)

Solutions to Worked Examples

A1

The half-life of a radioactive sample is the time required for one half of the nuclei in the sample to disintegrate, i.e. time for its activity to reduce to half its initial value.
(a) Half-life of the alpha source is 1.0 h from the graph.

(b) To get information about the beta source on its own, we would have to subtract the alpha activity at any time from the total activity at that time. Once this has been done, we can plot a half-life curve which refers only to the beta source. When this is done, the half-life turns out to be 6.0 h.

(c) Starting at time t = 0 h on the graph, we find that the combined source takes approx. 2.2 h to drop from 400 c.p.s. to 200 c.p.s. Then it takes approx. 3.8 h to drop from 200 c.p.s. to 100 c.p.s. After that, it takes about 6.0 h to drop from 100 c.p.s. to 50 c.p.s. Subsequent half-lives turn out to be 6.0 h, obviously because the alpha source is effectively 'dead' and the activity is now due only to the longer-lived beta source.

A2
(a) particle (b) electromagnetic waves (c) 5% speed of light
(d) speed of light (e) -1 unit (f) zero (g) 4 units
(h) $\frac{1}{1850}$ unit (i) a few centimetres (j) much greater than alpha
(k) a few mm of aluminium (l) several cm of lead (m) weak ioniser
(n) weak ioniser (o) small deflection (p) no deflection

A3
(a) The alpha particles are deflected (scattered) by the repulsion between them and the positively charged nuclei of the gold atoms in the foil.

(b) With the chamber evacuated, we can be certain that any deflection has been caused within the foil. At normal air pressure there is the possibility of collisions between the alpha particles and the molecules of the air. Such collisions would make the results more difficult to interpret.

(c) Rutherford proposed that atoms were largely empty space and therefore most of the alpha particles from the source would not be deflected much. He concluded that only powerful electrical repulsions could bring about the large angle deflections which were *occasionally* observed (see graph). His model of the atom, therefore, included a nucleus which contained the bulk of the mass of the atom and all the positive charge was concentrated in this nucleus. The nucleus occupies only a small fraction of the volume of the entire atom and so the chances of a particle coming near enough to the nucleus to suffer a large angle deflection are quite small. Most alpha particles suffer only small deflections.

(d) (i) This is a case of nuclear collisions between helium nuclei (α- particles) and nuclei of gold atoms. Rutherford suggested that large angle deflections could be caused by a single close encounter between an α- particle and the nucleus of the gold atom.
(ii) The chance of a close encounter is small since the atom is mostly space.

A4
(a) (i) With a negative charge on the electroscope, electrons emitted by ultra-violet radiation from the zinc plate would lower the charge on the plate and hence the p.d. between the leaf and the case of the electroscope. The leaf would therefore fall. With a positive charge on the electroscope electrons emitted from the zinc plate by u.v. light would not escape and so the divergence of the leaf would stay the same.
(ii) It is found by experiment that a clean zinc plate will release photoelectrons whereas steel, for example, will not. The negative zinc plate would in turn emit fewer photoelectrons than a caesium-coated cathode.
(iii) The frequency of the incident radiation is critical in determining whether enough energy is available to release electrons from the surface atoms of the metal plate. $E = hf_o$, where f_o is the frequency of radiation just needed to release electrons (the threshold frequency) and h is a constant.

(b) (i) Caesium is a better emitter of photoelectrons than zinc.
(ii) The frequency of the radiation (the colour of the light) and the intensity

86   Atomic physics

increase the reading.
(iii) Graph I indicates that a minimum threshold frequency is needed to initiate photoemission. After that, higher frequencies increase the current. Graph II indicates that, provided photoemission is possible, an increase in intensity produces an increase in current.

Practice Questions

1. (A1)
Radium (Ra 226) has a half-life of approximately 1.6×10^3 years.
(a) How long will it take for the Ra 226 to completely decay?
(b) How much of an active 1 kg sample will still be active after
   (i) 8.0×10^3 years?
   (ii) 10 half-life periods have elapsed?

2. (A1)
The corrected count-rates recorded during a half-life experiment are shown in the table below.

Corrected count-rate (c.p.s.)	200	50	1.6	0.8
Time (hours)	0	2	7	8

(a) What is the half-life of the substance being investigated?
(b) What would be the corrected count-rate after
   (i) 3 hours?
   (ii) 5 hours?

3. (A2)
Here are three descriptions of nuclear radiations. Try to identify the radiations in each case.
(a) Carries a charge and is therefore deflected by magnetic fields. Is a good ionising agent and has a relatively low speed.
(b) Has a large range in air but is not deflected by magnetic fields.
(c) Causes Geiger-Müller tubes to activate a counter unit, and travels at 90% the speed of light.

4. (A3)
(a) What kind of radiation did Geiger and Marsden use in their scattering experiments?
(b) What important information was gained from the experiment concerning the structure of the atom?
(c) How do the number of back-scattered particles compare with the number of particles transmitted through the gold foil?

5. (A4)
A pupil places an exposure meter, taken from a camera, in front of a white light source.
(a) How would you expect the reading on the meter to depend upon
   (i) the distance of the white light source from the meter?
   (ii) the power of the bulb placed in the light source?
(b) If a series of colour filters were placed in front of the white light source, which filter would you expect to produce the greatest reading on the exposure meter?
   (i) blue filter (ii) yellow filter (iii) red filter
   Assume that the intensity of the transmitted light is the same for each colour filter.
(c) Why might an ultra-violet lamp, which produces very little visible light, cause the exposure meter reading to go right off scale?

Answers to practice questions

Ray optics

1. (a) refractive index (b) two media involved in the refraction (c) 1.41
2. (a) 35.3° (b) 48.6° (c) 1.23
3. 0.46°
4. (a) Water : 48.8°, diamond : 24.4° (b) for glass, $45° > i_c$
5. When ray enters prism its angle of refraction is 15.1°. It then hits the opposite side at 44.9° to the normal and so is totally internally reflected towards the third side. It emerges from this side at 23° to the normal.
6. (b) v = 30 cm
7. (a) f = 5.7 cm (b) virtual, erect, magnified 8 x, – 40 cm from lens.
8. (a) (i) u = 3 cm (ii) f = 2.99 cm
   (b) Lens shifted away from film slightly ; size of image is less than before, but still magnified.

Wave motion

1. (a) (i) $f = 1.0 \times 10^{10}$ Hz (ii) $v = 2.12 \times 10^8$ m s^{-1} λ = 2.12 cm
   (b) microwaves reflected by metal
2. (a) path difference = whole number of wavelengths
   (b) (i) fringe width (ii) wavelength (iii) slit to screen distance (iv) slit separation
   (c) 6.0×10^{-7} m, 2.00 m, 1.0×10^{-4} m
   (d) light of one frequency only
3. 6.25×10^{-7} m
4. (a) There is only one spectrum with the prism (line or continuous depending upon source used). No second or higher order spectra can be obtained.
   (b) Different frequencies are refracted by slightly different amounts by the prism and so the colours disperse into a spectrum typical of the source used.
5. (a) 4.9×10^{-7} m (b) 34.8°

Mechanics

1. (a) 5 rev. s^{-1} (b) 0.2 s (c) 5 f.p.s.
2. (a) 20 f.p.s. (b) 15.0 cm
3. (a) 6 m s^{-2} (b) 19.1 m s^{-1} (c) 3.3 s
4. (a) 1.0 m s^{-1} (b) 3.0 m s^{-1}
5. (a) down (b) 2.5 m s^{-2}, 0 m s^{-2}, – 5.0 m s^{-2} (c) 800 N
6. (a) 9000 N (b) 30° to horizontal.
7. (a) (i) 180 000 N (ii) 20 000 N (b) 80 000 N
8. (a) 30 m s^{-1} & 40 m s^{-1} (b) horizontal component (c) 0 m s^{-1}
9. (a) 320 m (b) 170 m s^{-1} at 28.1° to ground.
10. (a) 6.0 kg m s^{-1} (b) 25.0 J (c) 5.0 m s^{-1} (d) 80 N
11. (a) 150 m s^{-1} (b) 11.24 (25) J
12. (a) 2 – d, inelastic (b) 2.0 m s^{-1}, 15.1° W of N.
13. (a) 0.21 m s^{-1}, at about 126° to original direction of puck which travelled at 0.6 m s^{-1}.

14. (a) Velocity grows from 0 m s^{-1} uniformly to 6.0 m s^{-1} and hits the ground 0.6 s after release. It is in contact for 0.1 s during which its velocity is reversed. It leaves the ground at 5.0 m s^{-1} and takes a further 0.5 s to reduce to zero velocity.
    (b) 1.1 J
15. z_{max} = 51.25 m
16. (a) 400 N (b) 160 J (c) 8 kW
17. (a) (i) 10 N (ii) 4 N (b) 6 W

Heat and kinetic theory of gases

1. (a) 4.9 x 10^6 J (b) 6.7° C (c) 6.7 kg
2. (a) C (b) A (c) A (d) C (e) 9.2 x 10^5 J
3. (a) 20° C (b) 2.4 x 10^6 J kg^{-1}
4. (a) 2.68 x 10^5 Pa (b) 18.7%
5. (b) 6.4 x 10^{-18} N (c) 3.2 x 10^{-12} Pa
6. (a) 4.2 x 10^5 Pa (b) 824 m s^{-1}

Electricity

1. (a) (i) doubled (ii) halved (iii) decreased (iv) increased
   (b) (i) no change (ii) decreased (iii) decreased (iv) increased
2. (a) $C = \dfrac{kA}{d}$ (b) $C = \dfrac{Q}{V}$ (c) farads, coulombs and volts
   (d) (i) x 10^{-3} (ii) x 10^{-6} (iii) x 10^{-9} (iv) x 10^{-12}
3. (a) 4 μF (b) 1.2 x 10^{-6} C
   (c) Greater voltages would break down the dielectric.
4. (a) (i) 0 V (ii) 4 V (b) 4 x 10^{-6} A
   (c) (i) 4 V (ii) 0 V (d) 0 A
   (e) Graph shows usual exponential decay of charging current.
5. (a) (i) slower rate (ii) same rate (iii) slower rate
   (b) (i) 36.0 s (ii) 10 μA (iii) double the product RC
6. (a) (i) Graph shows exponential rise from 0 V towards 2.0 V.
   (ii) Graph shows exponential decay from 2.0 V towards 0 V.
   (b) (i) 2.0 V (ii) 4.0 x 10^{-6} C
7. (a) (i) Graph shows an exponential decay from 2.0 V towards 0 V.
   (ii) Graph shows an exponential decay from 2.0 V towards 0 V.
   (b) (i) 0 C (ii) 0 V
8. (a) 3.2 x 10^{-19} C (b) 2 (c) Drop moves upwards, soon acquiring a terminal velocity
   (d) Weight down is just balanced by air friction force and buoyancy force.
9. (a) 3 Ω, 6 Ω and 12 Ω in series : R = 21 Ω
   (b) 3 Ω, 6 Ω and 12 Ω in parallel : R = 1$\tfrac{5}{7}$ Ω
   (c) 3 Ω and 6 Ω in parallel, with 12 Ω in series : R = 14 Ω
   (d) 6 Ω and 12 Ω in parallel, with 3 Ω in series : R = 7 Ω
   (e) 12 Ω and 3 Ω in parallel, with 6 Ω in series : R = 8$\tfrac{2}{5}$ Ω
   (f) 3 Ω and 6 Ω in series, with 12 Ω in parallel : R = 5$\tfrac{1}{7}$ Ω
   (g) 6 Ω and 12 Ω in series, with 3 Ω in parallel : R = 2$\tfrac{4}{7}$ Ω
   (h) 12 Ω and 3 Ω in series, with 6 Ω in parallel : R = 4$\tfrac{2}{7}$ Ω
10. (a) 2 A (b) 2 A (c) 2 A (d) 12 A (e) 6 A (f) 4 A
    (g) 2 V (h) 4 V (i) 6 V (j) 12 V (k) 12 V (l) 12 V
    (m) 4 W (n) 8 W (o) 12 W (p) 144 W (q) 72 W (r) 48 W

Answers to practice questions 89

11. (a) Insert series multiplier of value 9995 Ω. (b) Insert parallel shunt of value 0.005 Ω.
12. (a) 6.0 V (b) 4.0 V (c) decreased
13. (a) 3.0 Ω (b) 3.75 Ω
14. 0.12 Ω
15. (a) The V–I graph is a straight line falling from (12.0 V, 0 A) to (2.0 V, 10.0 A).
(b) 0.2 Ω
16. (a) top branch ratio = lower branch ratio
(b) galvanometer (c) sensitive to small currents, centre-zeroed. (d) (i) no (ii) no
17. (a) 14.0 Ω (b) AT = 25 cm.
18. (a) 0.50 V m^{-1} (b) 1.90 V (c) 2.50 V (d) 0.083 A
(e) All balance lengths would be increased.
19. (a) (i) 24 (ii) 6 A (iii) 0.15 A (b) 0.017 A
20. (a) 20 000 W (b) 116 kV (c) 483 : 1 approx. (d) 1 : 30
21. (a) 8.0 × 10^6 m s^{-1} (b) 5.0 × 10^{16}
22. (a) 0.05 s (b) Five waves with alternate troughs missing. (c) half-wave rectified
(d) re-set to '0.5 ms/cm'. (e) amplitude increased.
23. (a) 100 000 J (b) 339 V
24. (a) (i) 14.1 V (ii) 28.2 V (b) vertical line 5.7 cm long
25. (a) 5.7 V (b) (i) 200 V (ii) 141 V
26. (a) bright, dim, dim, bright (b) bulb dims (c) ohms (d) Wind a large number of turns of thick copper wire onto a soft iron core.
27. (a) (i) current reduces (ii) See solution to problem E23.
(b) (i) 100 Ω (ii) 1000 Ω (c) 50 mA and 33.3 mA
28. (a) dim, bright, dim, bright (b) bulb dims (c) ohms
29. (a) (i) increases (ii) See solution to problem E25.
(b) (i) 100 Ω (ii) 10 Ω (c) See solution to problem E25.
(d) 200 mA and 300 mA.
30. (a) (i) current drops (ii) current drops (b) current rises both in (i) and (ii)
(c) (i) $i_L > i_C$ (ii) $i_L = i_C$ (iii) $i_L < i_C$
31. (a) and (b) reduce in both cases, (c) halves.

Atomic physics

1. (a) infinite time interval (b) (i) 31.3 g (ii) 0.98 g
2. (a) 1 hour (b) (i) 25 c.p.s (ii) 6.25 c.p.s.
3. (a) alpha (b) gamma (c) beta
4. (a) alpha (b) see solution for A3 (c) a small fraction, see graph A3.
5. (a) (i) reduces as distance increases (ii) proportional to power of bulb
(b) the blue filter (c) high frequency and so high energy photons.

Index to worked examples

Absorption of nuclear
   radiations A2
Acceleration M2-8, 10, 13, 14
   due to gravity M3-5, 7, 8, 13, 14, 16
Aerial E26
Air
   resistance E5
   table M12
Alpha
   particle A2, 3
   particle scattering A3
   source A1, 3
Alternating
   current E14, 16, 17, 19, 21-27
   voltage E14, 16, 17, 19, 21-27
Ammeter E7, 8, 16
   conversion to E7
Amplifier E20
Amplitude E20
Angle
   of incidence and refraction L1-4, W1
   of projection M7, 8
Angular magnification L8
Anode E18
Area
   overlap of capacitor E2
   under v-t graph M2, 13
Astronomical telescope L8
Atmospheric pressure M17
Atomic physics A1-4
Average velocity M2, 4

Back e.m.f. E15, 22, 23
Background radiation A1
Back-scattering A3
Balanced forces M5, 6
Barometer M17
Beta
   particle A2
   source A1
Boiling point H1, 2
Boyle's law H3
Braking
   distance M14
   force M14
Branching currents E6, 7, 10, 11, 15, 22
Brightness levels of bulbs E9, 15, 16, 21, 22, 24

Cable E17
Caesium cathode A4
Capacitance
   factors affecting E1, 2
   a.c. circuits E24-26
Capacitive reactance E25
Capacitor E1-4, 24-26
   equation E2

Cathode E18, A4
   ray oscilloscope (see oscilloscope)
Cell
   driver E10-13
   standard E12, 13
Change
   in momentum M10-12
   of state H1, 2
Charge
   storage E1-4
   transfer E1-4, 18
Charging
   capacitors E1-4, 24, 25
   current E3, 4, 24-26
Cineprojector E16
Circular motion M1
Clapper board E20
Coil E14, 15, 22, 23, 26
Collimator W4
Collision M9, 11-13, H3, 4, A3
   nuclear A3
   2-dimensions M12
Components
   of force M4, 16
   of velocity M7, 8
Condensing steam H2
Connected bodies M6
Conservation
   of energy M7, 12, 14
   of momentum M9, 11
Converging lens L5-8
Corrected count-rate A1
Coulomb E2, 4, 5, 18
Count-rate A1
Critical angle L3, 4
Cross-sectional area of wire E22

Deceleration M2, 6, 7, 13, 14
Density of gas H4, 5
Dielectric E1, 2, 24, 25
Diffraction grating W4
Diminished images L5
Diode E18, 19
   equation E18
Direct current E3-15, 18-20
Discharging capacitors E4, 24
Dispersion effects L2
Displacement M2
Distance M2-4, 14
Double slit experiment W2
Driver cell E10-13

Effective resistance E6, 11
Efficiency of transformers E16, 17
E.H.T. supply E5, 18
Elastic collisions M12
Electric
   charge E1-5

circuits E3, 4, 6-12, 15-19, 22-27
   field patterns E5
   lamps/bulbs E9, 15, 16, 21, 22, 24
Electrical
   power E16, 17
   resonance E26, 27
Electricity E1-27
Electromagnet E14, 15, 22
Electromagnetism E14-17, 22, 23, 26
Electron E1, 5, 18
Electronic charge E5, 18
Electroscope A4
Electrostatics E1-5
E.M.F. E8, 9, 12, 13
   back E15, 22
Energy M7, 9, 10, 12-16
   electrical E16-18
   heat H1-6
   kinetic M7, 9, 10, 12-14 E18
   loss M9, 13
   potential M7, 13, 14
Enlarger L7
Equations of motion M2-4, 7, 8, 10, 13
Erecting lens L8
Exponential curve E3
Exposure meter A4
External resistance E8, 9
Eyepiece L8

Farads E2, 4
Filament E18
Flash-rate M1
Focal length L5-8
Focus L5-8
Force M4-6, 10, 11, 15, 16
   diagram M5, 6, 16
   of friction M4, 6, 15, 16
Frequency W1-4, E19, 20, 22-27
   of oscillation M1
   response of capacitor in a.c. E24, 25
   response of coil (inductor) in a.c. E22, 23
   resonant E26
   response of photocell A4
Friction M4, 6, 15, 16
   compensation M4
Frictionless motion M12, 14
Fringe patterns W2, 3
Full scale deflection E7
Fusion, specific latent heat of H1

Galvanometer
   moving coil E7
   centre-zero E10-13
Gamma radiation A2

Index to worked examples 91

Gas
  laws H3–6
  pressure M17, H2–6
Geiger and Marsden experiment
  A3
General gas equation H3, 5, 6
Generator E17
Geometrical optics L1–8
Gold foil A3
Gradient M16
Graphs
  activity-time A1
  area under v–t M2, 13
  current-frequency E23, 25, 26, 27
  current-time E3, 15
  pressure–temperature H3
  pressure–volume H3
  reactance-frequency E23, 25
  Snell's law L1
  temperature-time H1
  velocity-time M2, 7, 13
  voltage-time E4, 19, 24
Gravity M3–8, 13–17

Half life
  radiation A1
  R–C circuit E3
Half-wave rectification E19
Heat
  energy H1, 6
  losses E17, 26
Heater filament E18
Heating effect E21
Hoist M5
Horizontal
  component of velocity M7, 8
  force M6, 10, 11

Ideal gas H3–6
Image
  distance L5–8
  formation L5–8
Immersion heater H1
Impacts M9, 11, 12, 13
  in two dimensions M12
Impulse M10, 11
Inclined plane M3, 4, 16
Inductance E15, 26
Induction
  electromagnetic E14–17, 22, 23
  of charge E1
Inductive reactance E22, 23
Inductor E15, 22, 23, 26
Inelastic collision M9, 11, 13
Intensity of light W2, 4, A4
Interference
  of light W2
  of sound waves W3
Internal resistance E8, 9, 13
Inverted images L5–8
Ionisation by nuclear radiations A2

Joule (see energy)

Kelvin temperature H3, 5, 6
Kinetic
  energy M7, 9, 10, 12–14, E18

energy, average of molecules H5
theory of gases H3–6

Laser L4
L–C resonance E26
Lens
  converging (convex) L5–8
  equation L6, 7
  erecting L8
  eyepiece L8
  objective L8
Light L1–8, W2, 4, A4
  intensity E15, 21, 22, 24, A4
Lift in motion M5
Linear
  air track M11
  magnification L6, 7
Loss of energy M9, 13
Lost volts E8, 9
Loudspeaker W3, E14

Magnetic
  deflection of radiation A2
  field E15
Magnetism E14
Magnets E14
Magnification
  angular L8
  linear L5–8
Magnified images L5–8
Manometer M17
Mass M4–6, 9–14, 16, H1, 2, 4–6
  of electron E18
Mechanical
  energy M13
  power M15, 16
Mechanics M1–17
Melting point H1
Mercury M17
  vapour lamp A4
Metre-bridge E11
Microammeter E18
Microfarad E2, 4
Microphone W3, E20
Milliammeter E23, 25
Millikan's oil drop experiment E5
Momentum M9–12, H3, 4
  conservation of M9, 12
Monochromatic light W2, 4
Motor
  effect E14
  rule (right hand) E14
Multiflash photography M1
Multipliers and shunts E7

National grid system E17
Newton M4–6, 10, 11, 15–17
  balances M15
Newton's
  First Law of Motion M5, 6
  Second Law of Motion M4–6, 10
Nichrome wire E11–13
Normal L1–4, W1
Nuclear
  model of the atom A3
  physics A1–3
Null deflection methods E10–13

Object
  optical L5–8
  distance L5–8
Objective lens L8
Ohm E4, 6–13, 16–18, 23, 25
Ohm's Law E4–13, 16, 17
Optical centre L5, 6
Optics L1–8, W2, 4
Oscilloscope E19–21
Overlap area of capacitor E2, 25

Parallel
  combination of resistors E6, 11
  light L8, W4
  -plate capacitor E1, 2
  resonance E26
Pascal M17, H6
Path difference W2, 3
Peak
  current E23
  voltage E21, 22, 24
Pendulum M1
Period of oscillation M1
Phase difference W2, 3
Photocell E21, A4
Photoelectric effect A4
Photoelectron A4
Photographic enlarger L7
Plane water waves W1
Potential
  difference E1–13, 15–26
  divider E21
  energy M7, 13, 14
  gradient E12
Potentiometer E12, 13
Power
  electrical E16, 17
  heat H1
  loss E17
  mechanical M15, 16
  output M15, 16
Pressure M17
  of gas H3–6
Principle of conservation of momentum M9, 12
Prism L2
Projectile motion M7, 8
Protective resistor E11, 12
Puck M12

Radioactivity A1–3
Range
  of nuclear radiations A2
  of projectile M7, 8
Ray
  diagrams L1–8
  optics L1–8
R–C charging circuits E3, 4
Real images L5–8
Rectification E19
Refraction
  of light L1–8
  of water waves W1
Refractive index L1–4, W1
Resistance E3–13, 15–17, 19, 21, 26
Resistors
  damping effect E26
  protective E11, 12
  series and parallel E6
  standard E10, 11

Index to worked examples

Resolution of vectors M7, 8, 12, 16
Resonance, series and parallel E26
Resonant frequency E26
Resultant force M4-6, 10, 16
Rheostat E13
Right-hand motor rule E14
Root mean square
  current E23, 25
  speed H3-6
  voltage E21-25
Rubber ball M13
Rutherford's scattering experiment A3

Scattering of alpha particles A3
Scalars and vectors M1
Selectivity of resonant circuits E26
Self inductance E15
Semiconductor diode E19
Series
  combination of resistance E6
  resonance E26
Shunts and multipliers E7
Signal generators W3, E22, 24, 25
Simple pendulum M1
Slopes M3, 4, 16
Snell's law of refraction W1, L1-4
Sodium source W4
Soft iron core E14, 15, 22, 23
Specific
  heat capacity H1, 2
  latent heat of fusion H1
  latent heat of vaporisation H1, 2
Spectrometer W4
Spectrum L2, W4
Speed M1, 3, 5-14, 16
  of nuclear radiations A2
  of sound E20
  of water waves W1
Standard
  cell E12, 13
  resistor E10, 11
Standing waves E14

State, Change of H1, 2
Step-down transformer E15-17
Step-up transformer E17
Storing charge E1-4
S.T.P. M17
Stroboscope M1
Stroboscopic photography M1

Telescope L8, W4
Temperature H1-6
  -time graph H1
Tension M5, 6
Terminal
  potential difference E8, 9
  velocity E5
Terrestrial telescope L8
Thermionic
  diode E18
  emission E18
Thermometry H1-6
Thin lens equation L6, 7
Threshold frequency A4
Thrust M6
Time M1-16
  of flight M7, 8
  of impact M11
  -base setting E19, 20, 21
Total internal reflection L3, 4
Towing problem M6
Traces on oscilloscope E19-21
Transformer E15-17
Trolleys M4, 9
Tuned circuits E26
Tuning fork E20
Turns ratio E16, 17
Two-dimensional collisions M12

Ultra-violet radiation A4
Upright images L5, 8
Unbalanced forces M4-6
Uniform
  acceleration M2-8, 10, 13
  resistance wire E11-13, 21
  speed M1, 5, 6
  velocity M5, 6

Vacuum tube E18
Vaporisation, specific latent heat of H1, 2
Vector diagrams M7, 8, 12, 16
  of force M16
  of momentum M12
  of velocity M7, 8
Vector nature of momentum M9, 11, 12
Vectors and scalars M1
Velocity M1-14, 16
  -time graph M2, 7, 13
Vertical component of velocity M7, 8
Virtual images L5
Viscosity E5
Voltage (see potential difference)
Voltmeter E7-9
  conversion to E7
Volume of gas H3, 4

Water waves W1
Watts M15, 16
Wave
  equation E14
  interference W2, 3
Wavelength
  of light W2
  of sound W3
  of water waves W1
Wave optics W2, 4
Waves W1-4
  standing E14
Wheatstone bridge E10
Work M10, 14, 15, E18

Y-gain of oscilloscope E19-21
Young's double slit experiment W2

Zinc plate A4

© R. H. C. Neill, G. Sydserff 1981

First published 1981
by Edward Arnold (Publishers) Ltd.
41 Bedford Square, London WC1B 3DQ

ISBN 0 7131 0496 1

Neill, R. H. C.
  Upgrade Your Physics
  1. Physics – Problems, exercises, etc
  I. Title II. Sydserff, G.
  530'.076 QC32
  ISBN 0-7131-0496-1

All rights reserved. No part of this publication may be reproduced, stored in a retrieval system, or transmitted in any form or by any means, electronic, mechanical, photocopying, recording or otherwise, without the prior permission of Edward Arnold (Publishers) Ltd.

Printed in Great Britain by Spottiswoode Ballantyne Ltd.
Colchester and London